TESTING FOR CONTINUOUS DELIVERY

WITH VISUAL STUDIO 2012

Testing for Continuous Delivery with Visual Studio 2012

Larry Brader
Howie Hilliker
Alan Cameron Wills

978-1-62114-018-4

Contents

vi

x

Foreword

This book tells the story of two companies, Contoso and Fabrikam. Over my thirty years in the software industry, I've seen lots of companies that work like Contoso. Fortunately, over the last decade, I've also seen more and more like Fabrikam.

There never has been a better time for software. We can now produce a better flow of value for our customers, with less waste and more transparency than we ever could before. This revolution has been driven largely from the bottom, by small, agile, fast-moving development teams at places like Fabrikam. It's a global revolution, visible in places as unlike one another as Silicon Valley, Estonia, China, Brazil, and India.

Those of us in mature economies need to keep pace. Our supply chains are morphing into supply ecosystems. Software is no longer about designing everything to build; it's about finding parts to reuse and rapidly experimenting with innovation—the one thing you can't reuse or outsource. Every time you pick up your smartphone, you experience the benefits of this cycle.

Software testing needs to keep pace too. In the days of Contoso, we thought about minimizing "scrap and rework." Now at Fabrikam, we think of rapid experimentation and a continuous cycle of build-measure-learn. Testers are no longer the guys who catch stuff over the wall, but are full members of a multidisciplinary, self-organizing team. While their role at Contoso was defensive—do no harm to the customer, at Fabrikam it is offensive—anticipate the customer's behavior and wishes and be the first and best advocates.

Welcome to testing at Fabrikam. It's a lot more hectic around here, but it's also a lot more fun and rewarding.

Sam Guckenheimer
Redmond, Washington
May, 2012

Preface

Testing has always been the less glamorous sister to software development, scarcely noticed outside the software business, and the butt of complaints inside. If some small error is missed, testing is to blame; if time is running short or costs are getting too high, testing is liable to be starved of resources.

Testing certainly is expensive and time consuming. A project might easily spend 50% more on testing than on coding. Managers will quite reasonably ask whether the smallest update in the code really means that we must yet again configure all the machinery, call in the test team, search the office for the test plan, and work through a full suite of tests.

Back in the old days, testing came at the end of development, when the application's components were finally glued together. But that wasn't the end of the project, because a great many bugs would be found and there would then ensue a long period of repair, euphemistically called "stabilization." The end product of this process would be a full complement of last-minute patches and inconsistencies, and therefore difficult to update in the future.

Over the past decade or so, the software industry has been gradually learning to develop and test incrementally, integrating and testing continuously, and taking frequent feedback from customers. The applications that result from these agile processes are much more likely to satisfy their users.

But testing is still too often a huge, unwieldy process that acts as a roadblock, slowing the development cycle and limiting a development team's ability to respond to customer demands.

Today's software producers cannot afford such overhead. Web applications typically serve millions of people, and continuous bug-free operation is critical to customer satisfaction. These apps have to be updated frequently to keep up with shifting business needs, and at the same time have to maintain maximum up time for the health of the business. With so much riding on properly functioning and agile web apps, a faster develop, test, and deploy cycle is crucial.

In these modern applications, organizations cannot afford to implement long painful testing processes every time a feature needs tweaking or a line of code must change. That's why we wrote this book.

Fortunately, the continuous testing required today need not be as cumbersome as it was in the past. If you perform application lifecycle management using the Microsoft tools based on Visual Studio, the testing piece of your development puzzle will be less painful, time consuming, and expensive than it had always been. This combination of tools helps you automate testing, helps you track and record the process and results of testing, and helps you easily repeat that testing whenever you need to.

The key to this improved testing process is integration. Integrated tools mean that your test plan is a living and executable document that links directly to your tests. It means that you can take snapshots of your testing environment configuration for storage and later use and to memorialize the exact state of a system when a flaw was discovered. You can record tests for later playback and reuse—while you are performing a manual test, you can record the steps which will automatically play back next time you need to kick the tires on that app.

The first software systems wouldn't cause bodily injury if they made mistakes. But as computers found their way into cars, aircraft, hospital equipment, power stations, and factories, their destructive potential rose. Lives were on the line. The fault tolerance demanded by these applications soon became desirable in everyday applications. Many businesses today want or need such reliable outcomes, but they don't have the time or resources that the testing of old required.

It's a different world today in software development. We truly do operate on Internet time. The audience for our applications is huge, they're worldwide, and they don't sleep and allow us to roll out updates overnight. They click by the thousands and tens of thousands simultaneously, and they expect speed, availability, and precision. Today the business that can achieve 99.999 percent uptime, roll out updates continuously, and fix errors as soon as they are discovered will come out on top.

Fortunately, this set of tools, integrated so well with Visual Studio, will help you achieve that rapid fix-and-deploy goal and help keep you competitive. The team that brings you this book hopes, and feels confident that you will make better use of resources and have a much smoother test and release process. You just need to understand how application lifecycle management works in Visual Studio and begin testing the waters.

There are several other excellent books about software testing in Visual Studio, listed in the Bibliography on MSDN. This book provides a slightly different perspective, in the following respects:

- This is about how to use the Visual Studio tools effectively as a tester to deal with your product's entire lifecycle. In fact, our suite of lifecycle management tools includes Visual Studio, Microsoft Test Manager, and Team Foundation Server. If your job is to perform end-to-end tests of a complex system, this is one of the most powerful sets of integrated tools you can get. We'll show you how to use it not just to verify the basic functionality, but also to layer your test plans over multiple quality gates to achieve high fault tolerance.

- We think you'll take on this lifecycle management one piece at a time. That's normal. You probably already have some testing tools. It takes time to learn new tools, and even longer for the team to agree how to use them. Fortunately, you don't have to implement all of our advice all in one big bite. We've structured the book along the adoption path we suggest so as to make it easier to adopt it piece by piece.

- ALM—application lifecycle management—is changing. Increasingly, we're writing applications for the cloud. We need to publish updates every few weeks, not every few years. Applications aren't limited to the desktop anymore; a typical requirement might be implemented by an app on your phone and several collaborating servers on the web. Fault tolerance is a critical issue for 24x7 operations. It is no longer acceptable to leave system testing to the end of the project; testing needs to happen continuously. We'll show you how to use the methodology in this book to make this work.

- We describe how to set up and configure the machinery for testing. If your job is to administer Visual Studio, lifecycle management, and its testing tools, the Appendix is for you.

- We try to show you how to drive, not just operate the controls. The bibliography at the end includes some very good books and websites on topics like unit testing, test-driven development, model-driven testing, exploratory testing, and so on. We can't cover all that in this book, but we do take some of the most valuable patterns and show how you can use them to good effect.

We'll assume you have done some testing, or maybe a lot. You might be a test lead, or a developer on a team where developers also test, or a specialist tester. There are a few sections where you'll find it easier if you're the sort of tester who also writes some code; but that's not an essential prerequisite.

Testing is a highly skilled job. In addition to the constructive creativity of coding, it requires an intuition and depth of understanding that can find just those cases that will either find a bug or provide the confidence that the application works for many neighboring cases. It requires the application of many different techniques and categories of test, such as functional, load, stress, and security testing.

Like any job, there are also less skilled and more tedious parts—setting up machines, retesting the same feature you tested last week and the week before. It's these aspects that we hope you'll be able to mitigate by using our tools.

If you test software, you're the guardian of quality for your product. You're the one who makes sure it satisfies the customers. From that point of view, you're the real hero of the development team.

Unfortunately, we can't make software testing sound any more glamorous outside the industry. Bear this in mind when introducing yourself at social occasions. And if you have any good tips for that scenario, we'd love to hear from you.

The team who brought you this guide

Larry Brader conceived, motivated, and led the creation of this book. "I have been testing for a long time."

Chris Burns drew the technical illustrations.

Paul Carew drew the cartoons. He is a graphic designer who has worked for Microsoft and other major software companies in the Seattle, Washington area.

RoAnn Corbisier is a senior technical editor in Microsoft's Developer User Education team.

Nelly Delgado is our production editor.

Howard F. Hilliker wrote Chapter 6 and a lot of the testing content in the MSDN library, and provided support throughout. He has probably been Microsoft's tallest programmer and writer for fourteen years. "It's good to be writing a flowing storyline as opposed to reference pages."

Poornimma Kaliappan created the sample projects and logged the testing activities on which we based the examples in the book. "I have been a software development engineer in test with Microsoft for several years. This book has been an enjoyable trip."

Nancy Michell edited the text and kept us coherent.

Alan Cameron Wills devised the cartoons and wrote most of the words, including some brilliantly funny bits that were edited out.

We had a panel of expert reviewers who saved us from a number of embarrassing pitfalls. Thanks to Tim Elhajj (Microsoft Corporation), Katrina Lyon-Smith (Microsoft Corporation), Willy-Peter Schaub (Microsoft Corporation), Anutthara Bharadwaj (Microsoft Corporation), Muthukumaran Kasiviswanathan (Microsoft Corporation), Tiago Pascoal (Agilior), Paulo Morgado (Individual), Debra Forsyth (Object Sharp), Carlos dos Santos (CDS Informática Ltda.), Richard Hundhausen (Accentient), Paul Glavich (Saaus.com), Mike Douglas (Deliveron), Jacob Barna (Blue Cross and Blue Shield of Nebraska) and Marcelo Hideaki Azuma (ITGroup).

Thanks to Katie Niemer who invented our principal cartoon characters.

Thanks to Hans Bjordahl for the cartoon at the end. It's from his wonderful strip about life on a development team. See *http://www.bugbash.net*. Make sure you've got an hour or two to spare, and then click the "First" arrow when you get there.

Finally, thanks are due to the development team of Microsoft Test Manager, who created a wonderfully effective and well-integrated testing toolkit.

Where to go for more information

There are a number of resources listed in text throughout the book. These resources will provide additional background, bring you up to speed on various technologies, and so forth. For your convenience, there is a bibliography on MSDN that contains all the links so that these resources are just a click away: *http://msdn.microsoft.com/en-us/library/jj159339.aspx*.

You can also find this book online on MSDN: *http://msdn.microsoft.com/en-us/library/jj159345*.

1

The Old Way and the New Way

Today, software must meet your customers' needs in an ever-changing landscape. New features must be released continuously, updates must be timely, and bug fixes cannot wait for version 2. That's the reality in software development today, and it is never going back to the days of a new release every few years. That's especially true for cloud applications, but it's true for software in general as well.

Such agility requires lifecycle management that's built specifically to meet modern needs. But there are obstacles. Testing is one aspect of software development that can present a roadblock to agility. To publish even a one-line fix, you typically need to re-test the whole system. That can mean finding and wiring up the hardware, reassembling the team of testers, reinstalling the right operating systems and databases, and setting up the test harness. In the past, this overhead prevented the continuous update cycle that is so necessary today.

Many teams are still operating that old way, putting their competitiveness at risk. Fortunately, yours does not have to be such a team. You don't have to build and test software the way your ancestors did.

So how can you move your software testing into the 21st century? That is the subject of this book. This book is about how to streamline the testing of your software so that you can make updates and fix bugs more rapidly, and continuously deliver better software.

Because testing can be a significant obstacle, our aim in this book is to help you substantially reduce the overhead of testing while improving the reliability and repeatability of your tests. The result should be a shorter cycle of recognizing the need for a change, making the change, performing the tests, and publishing the update.

The planning and execution of this cycle is known as application lifecycle management (ALM).

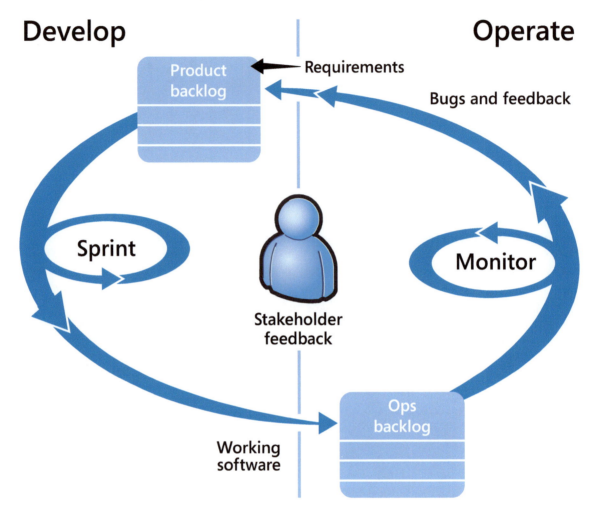

Develop

Operate

Product backlog

Requirements

Bugs and feedback

Sprint

Monitor

Stakeholder feedback

Ops backlog

Working software

Application lifecycle management with Visual Studio

Microsoft Visual Studio will figure prominently in this book and in your efforts to refine your own application lifecycle management. Application lifecycle management with Visual Studio is an approach that takes advantage of a number of Microsoft tools, most of which are found in Visual Studio and Team Foundation Server, and which support each part of the lifecycle management process. In these pages, we'll show you how to perform testing with Visual Studio Team Foundation Server 2012. Though we will focus on that version, you can also use Visual Studio 2010. (We note the differences where necessary.) With the setup guidance we provide here, your team should be able to test complex distributed systems within a few days.

Of course, you may already be doing everything we suggest here, so we can't promise the scale of improvements you'll see. But we envisage that, like us, you're concerned with the increasing demand for rapid turnaround. If your software runs in the cloud—that is, if users access it on the web—then you will probably want no more than a few days to pass between a bug report and its fix. The same is true of many in-house applications; for example, many financial traders expect tools to be updated within the hour of making a request. Even if your software is a desktop application or a phone app, you'll want to publish regular updates.

Before we dig into lifecycle management with Visual Studio, let's take a look at the two fictitious companies that will introduce us to two rather different approaches to testing and deployment: Contoso, the traditional organization, and Fabrikam, the more modern company. They will help define the ALM problem with greater clarity.

Contoso and Fabrikam; or something old, something new

At Contoso, testing has always been high priority. Contoso has been known for the quality of its software products for several decades. Recently however, testing has begun to seem like a dead weight that is holding the company back. When customers report bugs to Contoso, product managers will typically remark "Well, the fix is easy to code, but we'd have to re-test the whole product. We can't do that for just one bug." So they have to politely thank the customer for her feedback and suggest a workaround.

Meanwhile, their competitor, Fabrikam, a much younger company with more up-to-date methods, frequently releases updates, often without their users noticing. Bugs are gone almost as soon as they are reported. How do they do it? How do they test their whole product in such a short time?

In our story, the problems begin when Fabrikam and Contoso merge. The two cultures have a lot to learn from each other.

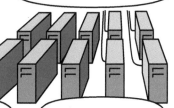

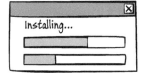

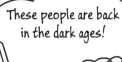

The two companies take very different approaches, and yours might be somewhere between the extremes. In the chapters that follow, we show you how to get set up to do things the new way, and we explain the choices you can make based on the situation in your organization.

From Contoso to Fabrikam

Let's take a look at the pain that Contoso experiences in the beginning and the benefits they realize as they move to a testing process more like Fabrikam's. You'll get a better understanding of the benefits to testing using Visual Studio, Team Foundation Server, and a virtual lab environment.

Here are some of the pain points that Contoso experienced by doing things the old way:

- Updating an existing product is expensive no matter how small the change. Partly, this is because test hardware has to be assigned, private networks have to be wired up, and operating system and other platform software has to be installed. This is a lengthy process. As a result, if customers find bugs in the operational product, they don't expect to see them fixed soon.

- During a project, the team frequently needs to test a new build. But the results are sometimes inconsistent because the previous build didn't uninstall properly. The most reliable solution would be to format the disk and install everything from scratch; but this is too costly.

- Manual tests sometimes yield inconsistent results. The same feature is tested by slightly different procedures on different occasions, even if the same person is doing the test.

- Developers often complain they can't reproduce a bug that was seen during a test run. Even if the tester has faithfully reported the steps that reproduce the bug, the conditions on a development machine might be different from those on the test environment.

- When the requirements change, it can be difficult to discover which tests should be updated; and it can be difficult to find out how well the latest build of the product meets the stakeholders' needs. Different unintegrated tools mean that there are no traceable relationships between bugs, tests, customer requirements, and versions of the product. The manual steps needed to make the tools work together make the process unreliable.

- Testing at the system level is always manual. To repeat a set of tests is costly, so that it is not economical to develop the software by incremental improvements. System testing is often abbreviated, so that bugs can go undiscovered until release.

- Revisiting code is risky and expensive. Changing code that has already been developed means rerunning tests, or running the risk that the changes have introduced new bugs. Development managers prefer to develop all of one part of the code, and then another, integrating the parts only towards the end of the project when it is often too late to fix integration problems.

Here are the benefits Fabrikam enjoys and which Contoso will realize by moving to the new way:

- Virtual machines are used to perform most tests, so new testing environments can be easily created and existing environments can rapidly be reset to a fresh state.

- Configurations of machines and software can be saved for future use: no more painful rebuilding process.

- Manual tests are guided by scripts displayed at the side of the screen while the tester works. This makes the tests more repeatable and reliable. Test plans and test scripts are stored in the Team Foundation Server database, and linked to the code and to the test configuration.

- Tests are easy to reproduce. Testers' comments, screenshots and actions can be recorded along with the test results. When a bug is found, a snapshot of the exact state of the environment is stored along with the bug. Developers can log in to the environment to investigate.
- Much more test automation. Manual test steps can be recorded and replayed rapidly. The recordings can also form the basis of automated system tests. Automated tests can be performed frequently, and at little cost.
- Reports on the project website show the progress of requirements in terms of the relevant tests.
- When requirements change, it's easy to trace which tests should be updated. An integrated set of tools manages and reports on requirements, test cases and test results, project planning, and bug management.
- In addition to the core tools, third-party products can be integrated.
- Changing existing code is commonplace. Automated tests are performed frequently and can be relied on to pick up bugs that were inadvertently introduced during a change. Development teams can produce a very basic version of an end-to-end working product at an early stage, and then gradually improve the features. This substantially improves the chances of delivering an effective product.

APPLICATION LIFECYCLE MANAGEMENT TOOLS

We'll assume that you've met Visual Studio, the Microsoft software development environment. As well as editing code in Visual Studio, you can run unit tests, break into the code to debug it, and use the IntelliTrace feature of Visual Studio to trace calls within the running code. You can analyze existing code with dependency, sequence, and class diagrams, and create models to help design or generate code.

Team Foundation Server is a tool for tracking and reporting the progress of your project's work. It can also be the source control server where your developers keep their code. It can build and test your software frequently as it grows, and provide reports and dashboards that show progress. In particular, you can get reports that show how far you've gone towards meeting the requirements—both in terms of work completed and tests passing.

Microsoft Test Manager (MTM) is the tool for testers. With it, you can plan and execute both manual and automated tests. While you are performing tests, you can log a bug with one click; the bug report will contain a trace of your recent actions, a snapshot of the state of the system, and a copy of any notes you made while exploring the system. You can record your actions in the test case, so that they can be played back on later occasions.

MTM also includes tools for setting up and managing lab machines. You can configure a virtual lab in which to install a distributed system, and link that lab to the test plan. Whenever you need to repeat tests—for example when you want to publish a change to your system—the lab can be reconfigured automatically.

Moving to the new way: adopting Visual Studio for ALM

Now we aren't suggesting that everyone should work by the same methods. Some teams want to aim for a rapid development cycle; others are developing embedded software for which that wouldn't be appropriate.

Nor do we suggest that people in Contoso are doing it all wrong. Their attitude to testing is clearly admirable: they take a pride in releasing high-quality software.

But we do believe that, no matter what development books a team reads, they do need to run tests, and they can benefit from having a brisker turnaround in their test runs—if only to reduce the boredom of repeating the same old tests by hand.

But such changes aren't just about adopting tools. A software team—the extended team that includes all the stakeholders who collaborate to produce the working software—consists of interacting individuals. To keep everyone in sync, new ways of doing things have to be tried out and agreed upon in measured steps.

Visual Studio provides a lot of different facilities. It therefore makes sense to adopt it one step at a time. Of course, this book is about testing, but since the test tools and other features such as work tracking and source control are closely integrated, we have to talk about them to some extent too.

The following diagram shows a typical order in which teams adopt Visual Studio for application lifecycle management. They begin by just using Visual Studio, and work gradually on up through source control, server builds, system testing, and on to automated system tests. The need for gradual progress is mostly about learning. As your team starts to use each feature, you'll work out how to use it in the best way for your project.

And of course a team learns more slowly than any of its members. Learning how to open, assign, and close a bug is easy; agreeing who should do what takes longer.

Adoption Maturity

Automated build, deploy and test for distributed systems

Lab environments
Test distributed systems.
Clean configuration on virtual machines.

Microsoft Test Manager
Plan and run repeatable tests.

Build service
Automated integration tests:
continuous, regular, or on demand.

Project portal
SharePoint site for project documents.
Project progress reports.

Team Foundation Server basics

Source control
Check out, check in.

Task management
Task and bug lists.

Just Visual Studio
Individuals use Visual Studio for development.

... and unit testing.

There are a number of ways you can gradually adopt Visual Studio for ALM, and the steps below represent one way. Naturally, this isn't a precise scheme. But here's what you get in each of the stages of adoption that we've shown:

- **Just Visual Studio** – Visual Studio is used for development. To test the application, Contoso's testers basically just press F5 and find out whether the application works.
  Unit tests are written by the developers to test individual classes and components.
  Coded UI Tests are a neat way to run automated tests of the whole application through its user interface.
- **Team Foundation Server Basics** – when you install Team Foundation Server, you get a host of features. The first features you'll want to take advantage of are:
  - **Source Control** to avoid overwriting each other's work. And after a while, you might start using:
    Check-in rules – which remind developers to run tests and quality analysis on their code before checking it into the server.
    Shelvesets – a way of copying a set of changes from one user to another for review before checking in.
    Branches – which help manage work in a large project.
  - **Task Management** is about tracking the backlog (the list of tasks), bugs, issues, and requirements. Each item is recorded in a work item. You can assign work items to people, project iterations, and areas of work; you can organize them into hierarchies; and you can edit them in Excel, and sync with other project management tools.
- The **Project Portal** is a SharePoint website, integrated with Team Foundation Server so that each project automatically gets its own site. And, even more interestingly, you can get some very nice dashboards, including graphs and charts of your project's progress. These reports are based on the work items, and on the test results.
- The **Build Service** is a feature of Team Foundation Server that performs a vital function for the development team. It builds all the code that has been checked in by developers, and runs tests. Builds can run on a regular or continuous cycle, or on demand. The team gets email alerts if a compilation or test fails, and the project portal shows reports of the latest results.
  The email alert is very effective at keeping code quality high: it prominently mentions who checked in code before the failed build.
- **Microsoft Test Manager** is where it gets interesting from the point of view of the professional tester. Microsoft Test Manager makes tests reliably repeatable and speeds up testing. Using it, you can:
  - Write a script for each manual test, which is displayed at the side of the screen while the test is being performed.
  - Partially automate a test by recording the test actions as you perform them. The next time you run the test, you can replay the actions.
  - Fully automate a test so that it can run in the build service. To do this, you can adapt code generated from partly automated tests.
  - Associate tests with user requirements. The project portal will include charts that display the progress of each user requirement in terms of its tests.

- Organize your tests into suites and plans, and divide them up by functional areas and project iterations.
- Perform one-click bug reporting, which includes snapshots of the state of the machine.
- **Lab Environments** are collections of test machines—particularly virtual machines. Without a lab, you can test an application locally, running it on your own computer. During development, applications are typically debugged on the development machine, often with several tiers running on the same machine. But with lab facilities, you can:
  - Deploy a system to one or more machines and collect test data from each machine. For example, a web client, Microsoft Internet Information Services (IIS), and a database would run on separate machines.
  - Run on freshly-created virtual machines, so that there's no need to uninstall old versions, no chance of the application corrupting your own computer, and you can choose any platform configuration you like.
  - Configure an environment of virtual machines for a particular test suite, and store it for use whenever you want to run that suite again.
  - Take a snapshot of the state of an environment and save it along with a bug report.
- **Automated build, deploy, and test**. The simplest setup of the build service runs unit tests in the same way the developer typically does—all on one machine. But for web and other distributed applications, this doesn't properly simulate the real operational conditions. With automated deployment, you can run tests on a lab environment as part of the continuous or regular build. The automation builds the system, instantiates the appropriate virtual environment for the tests, deploys each component to the correct machine in the environment, runs the tests, collects data from each machine, and logs the results for reporting on the project portal.Now let's take a look at what you'll find in the remaining chapters.

Chapter 2: Unit Testing: Testing the Inside

Developers create and run unit tests by using Visual Studio. These tests typically validate an individual method or class. Their primary purpose is to make sure changes don't introduce bugs. An agile process involves the reworking of existing software, so you need unit tests to keep things stable.

Typically developers spend 50 percent of their time writing tests. Yes, that is a lot. The effort is repaid many times over in reduced bug counts. Ask anyone who's tried it properly. They don't go back to the old ways.

Developers run these tests on their own machines initially, but check both software and tests into the source control system. There, the build service periodically builds the checked-in software and runs the tests. Alarms are raised if any test fails. This is a very effective method of ensuring that the software remains free of bugs—or at least free of the bugs that would be caught by the tests. It's part of the procedure that when you find a bug, you start by adding new tests.

Chapter 3: Lab Environments

To test a system, you must first install it on a suitable machine or set of machines. Ideally, they should be fresh installations, starting from the blank disc because any state lingering from previous installations can invalidate the tests. In Visual Studio, lab environments take a lot of the tedium out of setting up fresh computers and configuring them for testing.

A lab environment is a group of computers that can be managed as a single entity for the purposes of deployment and testing. Typically the computers are virtual machines, so you can take snapshots of the state of the complete environment and restore it to an earlier state. Setting up a new environment can be done very quickly by replicating a template.

Chapter 4: Manual System Tests

System tests make sure that the software you are developing meets the needs of the stakeholders. System tests look at what you can do and see from outside of the system: that is, from the point of view of users and other systems that are external to yours.

In many organizations, this kind of testing is done by specialist testers who are not the same people as the developers. That's a strategy we recommend. A good tester can write software and a good developer can test it. But you don't often find the strongest skills of creating beautiful software coexisting in the same head as the passion and cunning that is needed to find ingenious ways to break it.

System testing is performed with Microsoft Test Manager. As well as planning tests and linking them to requirements, Microsoft Test Manager lets you set up lab environments—configurations of machines on which you run the tests.

While you are running tests, Microsoft Test Manager's Test Runner sits at the side of the screen, prompting you with the steps you have to perform. It lets you record the results and make notes, and will record the actions you take to help diagnose any bugs that you find. You can log a bug with one click, complete with a screenshot, a snapshot of the machine states, and a log of the actions you took leading up to the failure.

Chapter 5: Automated System Tests

System testing starts with exploration—just probing the system to see what it does and looking for vulnerabilities ad hoc. But gradually you progress to scripted manual testing, in which each test case is described as a specific series of steps that verifies a particular requirement. This makes the tests repeatable; different people can work through the same test, without a deep understanding of the requirement, and reliably obtain the same result.

Manual tests can be made faster by recording the actions of the first tester, and then replaying them for subsequent tests. In the later tests, the tester only has to verify the results of each step (and perform some actions that are not accurately recorded).

But the most effective tests are performed entirely automatically. Although it requires a little extra effort to achieve this, the payback comes when you run the tests every night. Typically you'll automate the most crucial tests, and leave some of the others manual. You'll also continue to do exploratory manual testing of new features as they are developed. The idea is that the more mature tests get automated.

A fully automated system test builds the system, initializes a lab environment of one or more machines, and deploys the system components onto the machines. It then runs the tests and collects diagnostic data. Bug reports can be logged automatically in the case of failures. Team members can view results on the project website in terms of the progress of each requirement's tests.

Chapter 6: A Testing Toolbox

Functional tests are just the beginning. You'll want to do load tests to see if the system can handle high volumes of work fast enough; stress tests to see if it fails when short of memory or other resources; as well as security, robustness, and a variety of other kinds of tests.

Visual Studio has specialized tools for some of these test types, and in others there are testing patterns we can recommend.

Discovering a failure is just the first step to fixing the bug. We have a number of tools and techniques that help you diagnose the fault. One of the most powerful is the ability to save the state of the lab machines on which the failure occurred, so that the developer can log in and work out what happened. Diagnostic data adapters collect a variety of information while the tests are running, and IntelliTrace records where the code execution went prior to the failure.

Lab environments can be run by developers from Visual Studio while debugging—they aren't just a tool for the system tester.

Chapter 7: Testing in the Software Lifecycle

Whether you're aiming for deployment ten times a day, or whether you just want to reduce the cost of running tests, it isn't just a matter of running the tools. You have to have the right process in place.

Different processes are appropriate for different products and different teams. Continuous delivery might be appropriate for a social networking website, but not less so for medical support systems.

Whether your process is rapid-cycle or very formal, you can lower the risks and costs of software development by adopting some of the principles of agile development, including rigorous testing and incremental development. In this chapter we'll highlight the testing aspects of such processes: how testing fits into iterative development, how to deal with bugs, what to monitor, and how to deal with what you see in the reports.

Appendix: Setting up the Infrastructure

If you're administering your test framework, the Appendix is for you. We walk through the complete setup and discuss your options. If you follow it through, you'll be ready to hire a team and start work. (Alternatively, a team will be ready to hire you.)

We put this material at the end because it's quite likely that someone else has already done the setting up, so that you can dig right into testing. But you'll still find it useful to understand how the bits fit together.

The bits we install include: Visual Studio Team Foundation Server and its source and build services; Microsoft SharePoint Team Services, which provides the project website on which reports and dashboards appear; Microsoft Hyper-V technology and Microsoft System Center Virtual Machine Manager (SCVMM) to provide virtual machines on which most testing will be performed; lab management to manage groups of machines on which distributed systems can be tested; a population of virtual machine templates that team members will use; and a key server to let you create new copies of Windows easily. We'll also sort out the maze of cross-references and user permissions needed to let these components work together.

The development process

To simplify our book, we'll make some assumptions about the process your team uses to develop software. Your process might use different terms or might work somewhat differently, but you'll be able to adapt what we say about testing accordingly. We'll assume:

- Your team uses Visual Studio to develop code, and Team Foundation Server to manage source code.
- You also use Team Foundation Server to help track your work. You create work items (that is, records in the Team Foundation Server database) to represent requirements. You might call them product backlog items, user stories, features, or requirements. We will use the generic term "requirement." When each requirement has been completed, the work item is closed.
- You divide the schedule of your project into successive iterations, which each last a few weeks. You might call them sprints or milestones. In Team Foundation Server, you assign work items to iterations.
- Your team monitors the progress of its work by using the charts on the project website that Team Foundation Server provides. The charts are derived from the state of the work items, and show how much work has been done, and how much remains, both on the whole project and on the current iteration.
- You have read *Agile Software Engineering with Visual Studio* by Sam Guckenheimer and Neno Loje (Addison-Wesley Professional, 2011), which we strongly recommend. It explains good ways of doing all the above.

In this book, we will build on that foundation. We will recommend that you also record test cases in Team Foundation Server, to help you track not just what implementation work has been done, but also how successfully the requirements are being met.

TESTERS VS. DEVELOPERS?

In some software development shops, there's a deep divide between development and test. There are often good reasons for this. If you're developing an aircraft navigation system, having a test team that thinks through its tests completely independently from the development team is very good hygiene; it reduces the chances of the same mistaken assumptions propagating all the way from initial tender to fishing bits out of the sea. Similar thinking applies to the acceptance tests at the end of a traditional development contract: when considering whether to hand over the money, your client does not want the application to be tested by the development team.

Contoso operates a separate test team. When a product's requirements are determined, the test leads work out a test plan, setting out the manual steps that more junior testers will follow when an integrated build of the product becomes available.

This divide is less appropriate for Fabrikam's rapid cycle. Testing has to happen more or less concurrently with development. The skills of the exploratory manual tester are still required, but it is useful if, when the exploration is done, that person can code up an automated version of the same tests.

Contractual acceptance tests are less important in a rapid delivery cycle. The supplier is not done with the software as soon as it is delivered. Feedback will be gathered from the operational software, and when a customer finds a bug, it can be fixed within days.

These forces all lead towards a more narrow division between testers and developers. Indeed, many agile teams don't make that distinction at all. Testing of all kinds is done by the development team.

That isn't to say the separate approach is invalid. Far from it; where very high reliability is sought, there is a strong necessity for separate test teams. But there are also some companies like Contoso, in which separate test teams are maintained mostly for historical reasons. They could consider moving more towards the developers=testers end of the slider.

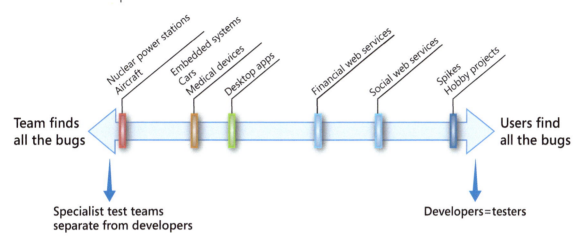

Who finds the bugs?

Where dev and test are separate, unit testing is the province of the developers, and whole-system testing is what the test team does. But even where there is a strong divide for good reasons, our experience is that it can be very useful to move people between the teams to some extent. Developers benefit from thinking about what's required in terms of exact tests, and testers benefit from understanding development. Knowing the code helps you find its vulnerabilities, and automating tests allows you to run tests more reliably and more often.

Agile development

We recommend that, if you aren't doing so already, you should consider using more agile principles of development. Agile development is made possible by a strong testing regime.

Please note, "agile" doesn't mean giving up a rigorous approach to project governance where that is appropriate. If you develop software for my car's braking system or my bank's accounting systems, then I would ask you to keep doing the audit trails, the careful specifications, and so on.

At the same time, it is true of any project—formal or not—that an iterative approach to project planning can minimize the risks of failed projects and increase the capacity to respond to changes in the users' needs.

To see why, consider a typical large development project. The Contoso team needs to develop a website that sells ice cream. (Okay, forget that they could just get an off-the-shelf sales system. It's an example.) A morning's debate determines all the nice features they would like the site to have, and the afternoon's discussion leads to a block diagram in which there is a product catalog database, a web server, an order fulfillment application, and various other components.

Now they come to think of the project plan. One of the more traditionally-minded team members proposes that they should start with one of the components, develop it fully with all the bells and whistles, and then develop the next component fully, and so on. Is this a good strategy? No. Only near the end of the project, when they come to sew all the parts together, will they discover whether the whole thing works properly or not; and whether the business model works; and whether their prospective customers really want to buy ice cream from the internet.

A better approach is to begin by developing a very basic end-to-end functionality. A user should be able to order an ice cream; no nice user interface, no ability to choose a flavor or boast on networking sites about what he is currently eating. That way, you can demonstrate the principle at an early stage, and maybe even run it in a limited business trial. The feedback from that will almost certainly improve your ideas of what's required. Maybe you'll discover it's vital to ask the users whether they have a refrigerator.

Then you can build up the system's behavior gradually: more features, better user interface, and so on.

But wait, objects a senior member of the team. Every time you add new behavior, you'll have to revisit and rework each component. And every time you rework code, you run the risk of introducing bugs! Far better, surely, to write each component, lock it down, and move on to the next?

No. As stated, you need to develop the functionality gradually to minimize the risk of project failure. And you'll plan it that way: each step is a new feature that the user can see. A great benefit is that everyone—business management, customers, and the team—can see how far the project has come: progress is visible with every new demonstrated feature.

But, to address that very valid objection, how do you avoid introducing bugs when you rework the code? Testing. That's what this book is about.

Traditional vs. Agile Development

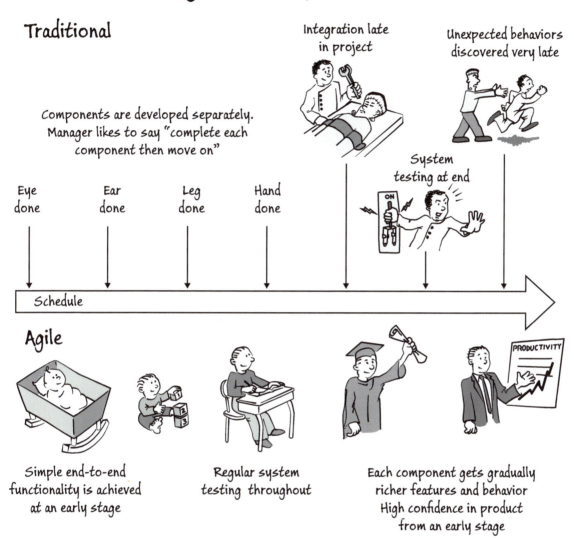

Traditional

Integration late in project

Unexpected behaviors discovered very late

Components are developed separately. Manager likes to say "complete each component then move on"

System testing at end

| Eye done | Ear done | Leg done | Hand done |

Schedule

Agile

Simple end-to-end functionality is achieved at an early stage

Regular system testing throughout

Each component gets gradually richer features and behavior. High confidence in product from an early stage

PRODUCTIVITY

Summary

Testing the whole system should no longer slow down your responsiveness when your software products must be updated. Provided you adapt your development and operational processes appropriately, there are tools that can help you automate many of your tests and speed up manual testing for the rest. The integration of testing tools with requirements and bug tracking makes the process reliable and predictable. The ability to save a test plan, complete with lab configurations, and to work through the test interactively, allows you to repeat tests quickly and reliably.

Where to go for more information

All links in this book are accessible from the book's online bibliography available on MSDN: *http://msdn.microsoft.com/en-us/library/jj159339.aspx.*

2

Unit Testing:
Testing the Inside

What drives Fabrikam's development process is the desire to quickly satisfy their customer's changing demands. As the more agile of the two companies in our story, Fabrikam's online users might see updates every couple of weeks or even days. Even in their more conventional projects, clients are invited to view new features at frequent intervals.

These rapid cycles please the customers, who like to see steady progress and bugs fixed quickly. They also greatly reduce the chances of delivering a product that doesn't quite meet the client's needs. At the same time, they improve the quality of the product.

At Contoso, the more traditional of the two, employees are suspicious of Fabrikam's rapid cycles. They like to deliver a high quality product, and testing is taken very seriously. Most of their testing is done manually, exercising each user function by following scripts: click this button, enter text here, verify the display there. It can't be repeated every night. They find it difficult to see how Fabrikam can release updates so rapidly and yet test properly.

But they can, thanks to automation. Fabrikam creates lots of tests that are written in program code. They test as early as possible, on the development machine, while the code is being developed. They test not only the features that the user can see from the outside, but also the individual methods and classes inside the application. In other words, they do unit testing.

Unit tests are very effective against regression—that is, functions that used to work but have been disturbed by some faulty update. Fabrikam's style of incremental development means that any piece of application code is likely to be revisited quite often. This carries the risk of regression, which is the principal reason that unit tests are so popular in Fabrikam and agile shops like them.

This is not to deny the value of manual testing. Manual tests are the most effective way to find new bugs that are not regressions. But the message that Fabrikam would convey to their Contoso colleagues is that by increasing the proportion of coded tests, you can cycle more rapidly and serve your customers better.

For this reason, although we have lots of good things to say about manual testing with Visual Studio, we're going to begin by looking at unit testing. Unit tests are the most straightforward type of automated test; you can run them standalone on your desktop by using Visual Studio.

> **Tip:** *Unit testing does involve coding. If your own speciality is testing without writing code, you might feel inclined to skim the rest of this chapter and move rapidly on to the next. But please stick with us for a while, because this material will help you become more adept at what you do.*

In this chapter

The topics for this chapter are:

* Unit testing on the development machines.
* Checking the application code and tests into the source code store.
* The build service, which performs build verification tests to make sure that the integrated code in the store compiles and runs, and also produces installer files that can be used for system tests.

Hey Art

Whaddya think about all this unit test stuff? I mean, it's all coding. Developers' stuff. I don't write code. I run the stuff and log bugs. Are we going to get rid of manual tests and just code everything?

I don't think so. Fabrikam still does manual testing, just like us. But they get more of the repetitive testing done by automating.

Yeah, unit testing means you don't have to worry about the obvious bugs that are easy to find. If we automate the easy tests, you can spend more time looking for interesting problems. Where we didn't understand what was needed, or didn't foresee some subtle interaction.

Or security vulnerablities...

Yeah! Stuff devs don't like to think about!

If you've a good feeling for where the users are coming from, you're a good tester. We're always going to need good testers. Unit tests free up your time so you can find the really interesting bugs - the bugs no-one ever expected. You'll be the hero of the team!

Maybe I'll be around for another year or two...

Prerequisites

To run unit tests, you need Visual Studio on your desktop.

The conventions by which you write and run unit tests are determined by the unit testing framework that you use. MSTest comes with Visual Studio, and so our examples use that.

But if you already use another unit testing framework such as NUnit, Visual Studio 2012 will recognize and run those tests just as easily, through a uniform user interface. It will even run them alongside tests written for MSTest and other frameworks. (Visual Studio 2010 needs an add-in for frameworks other than MSTest, and the integration isn't as good as in Visual Studio 2012.)

Later in this chapter, we'll talk about checking tests and code into the source repository, and having the tests run on your project's build service. For that you need access to a team project in Visual Studio Team Foundation Server, which is installed as described in the Appendix.

Unit tests in Visual Studio

A unit test is a method that invokes methods in the system under test and verifies the results. A unit test is usually written by a developer, who ideally writes the test either shortly before or not long after the code under test is written.

To create an MSTest unit test in Visual Studio, pull down the **Test** menu, choose **New Test**, and follow the wizard. This creates a test project (unless you already had one) and some skeleton test code. You can then edit the code to add tests:

```C#
[TestClass]
public class RooterTests
{
  [TestMethod] // This attribute identifies the method as a unit test.
  public void SignatureTest()
  {
    // Arrange: Create an instance to test:
    var rooter = new Rooter();

    // Act: Run the method under test:
    double result = rooter.SquareRoot(0.0);

    // Assert: Verify the result:
    Assert.AreEqual(0.0, result);
  }
}
```

Each test is represented by one test method. You can add as many test methods and classes as you like, and call them what you like. Each method that has a **[TestMethod]** attribute will be called by the unit test framework. You can of course include other methods that are called by test methods.

If a unit test finds a failure, it throws an exception that is logged by the unit test framework.

RUNNING UNIT TESTS

You can run unit tests directly from Visual Studio, and they will (by default) run on your desktop computer. (More information can be found in the MSDN topic *Running Unit Tests with Unit Test Explorer.*) Press CTRL+R, A to build the solution and run the unit tests. The results are displayed in the Test Explorer window in Visual Studio:

Unit test results

(The user interface is different in Visual Studio 2010, but the principles are the same.)

If a test fails, you can click the test result to see more detail. You can also run it again in debug mode.

The objective is to get all the tests to pass so that they all show green check marks.

When you have finished your changes, you check in both the application code and the unit tests. This means that everyone gets a copy of all the unit tests. Whenever you work on the application code, you run the relevant unit tests, whether they were written by you or your colleagues.

The checked-in unit tests are also run on a regular basis by the build verification service. If any test should fail, it raises the alarm by sending emails.

DEBUGGING UNIT TESTS

When you use **Run All**, the tests run without the debugger. This is preferable because the tests run more quickly that way, and you don't want passing tests to slow you down.

However, when a test fails, you might choose **Debug Selected Tests**. Don't forget that the tests might run in any order.

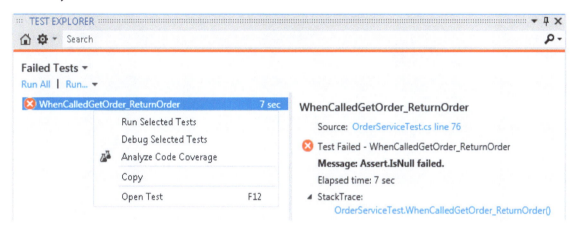

Running or debugging tests

Test-first development

Writing unit tests before you write the code—*test-first development*—is recommended by most developers who have seriously tried it. Writing the tests for a method or interface makes you think through exactly what you want it to do. It also helps you discuss with your colleagues what is required of this particular unit. Think of it as discussing samples of how your code will be used.

For example, Mort, a developer at Fabrikam, has taken on the task of writing a method deep in the implementation of an ice-cream vending website. This method is a utility, likely to be called from several other parts of the application. Julia and Lars will be writing some of those other components. They don't care very much how Mort's method works, so long as it produces the right results in a reasonable time.

What happens if we call it with a negative number?

But Julia, I'll document that the parameter only takes positive numbers. So it won't happen.

Tee hee!! Mort, this method will be around long after you've gone. It will be called by people you'll never meet. And they won't read your documents!

It should throw an exception.

Yeah, I guess. So OK. Here's an additional test.

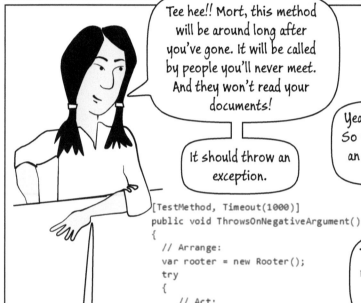

```
[TestMethod, Timeout(1000)]
public void ThrowsOnNegativeArgument()
{
    // Arrange:
    var rooter = new Rooter();
    try
    {
        // Act:
        rooter.SquareRoot(-2);

        // Assert: If we get here, we didn't throw:
        Assert.Fail();
    }
    catch (ArgumentOutOfRange)

    // Test passed - just return.
```

This test calls my same method, but exercises a different aspect of its behavior.

Yep. That's cool.

Cool. So we all know what my method should do. Now all I have to do is write some code that makes the tests pass!

Yeah. Just don't forget we might change our minds tomorrow.

Mort wants to make sure that he understands how people want to use his component, so he writes a little example and circulates it for comment. He reasons that although they aren't interested in his code, they do want to know how to call it. The example takes the form of a test method. His idea is that the test forms a precise way to discuss exactly what's needed.

Julia and Lars come back with some comments. Mort adjusts his test, and writes another to demonstrate different aspects of the behavior they expect.

Julia and Lars can be confident that they know what the method will do, and can get on with writing their own code. In a sense, the tests form a contract between the writer and users of a component.

Tip: *Think of unit tests as examples of how the method you're about to write will be used.*

Mort frequently writes tests before he writes a piece of code, even if his own code is the only user. He finds it helps him get clear in his mind what is needed.

LIMITATIONS OF TEST-FIRST DEVELOPMENT?

Test-first development is very effective in a wide variety of cases, particularly APIs and workflow elements where there's a clear input and output. But it can feel less practical in other cases; for example, to check the exact text of error reports. Are you really going to write an assertion like:

```C#
Assert.AreEqual("Error 1234: Illegal flavor selected.", errorMessage);
```

Baseline tests

A common strategy in this situation is to use the baseline test. For a baseline test, you write a test that logs the output of your application to a file. After the first run, you verify manually to see that it looks right. For subsequent runs, you write test code that compares the new output to the old log, and fails if anything has changed. It sounds straightforward, but typically you have to write a filter that allows for changes like time of day and so on.

Many times, a failure occurs just because of some innocuous change, and you get used to looking over the output, deciding there's no problem, and resetting the baseline file. Then on the sixth time it happens, you miss the crucial thing that's actually a bug; and from that point onwards, you have a buggy baseline. Use baseline tests with caution.

Tests verify facts about the application, not exact results

Keep in mind that a test doesn't have to verify the exact value of a result. Ask yourself what facts you know about the result. Write down these facts in the form of a test.

For example, let's say we're developing an encryption method. It's difficult to say exactly what the encrypted form of any message would be, so we can't in practice write a test like this:

```C#
string actualEncryption = Encrypt("hello");
string expectedEncryption = "JWOXV";
Assert.AreEqual(expectedEncryption, actualEncryption);
```

But wait. Here comes a tip:

> **Tip:** *Think of facts you know about the result you want to achieve. Write these as tests.*

What we can do is verify a number of separate required properties of the result, such as:

```csharp
C#
// Encryption followed by decryption should return the original:
Assert.AreEqual (plaintext, Decrypt(Encrypt(plaintext)));

// In this cipher, the encrypted text is as long as the original:
Assert.AreEqual (plaintext.Length, Encrypt(plaintext).Length);

// In this cipher, no character is encrypted to itself:
for(int i = 0; i < plaintext.Length; i++)
    Assert.AreNotEqual(plaintext[i], Encrypt(plaintext)[i]);
```

Using assertions like these, you can write tests first after all.

How to use unit tests

In addition to test-first (or at least, test-soon) development, our recommendations are:

- A development task isn't complete until all the unit tests pass.
- Expect to spend about the same amount of time writing unit tests as writing the code. The effort is repaid with much more stable code, with fewer bugs and less rework.
- A unit test represents a requirement on the unit you're testing. (We don't mean a requirement on the application as a whole here, just a requirement on this unit, which might be anything from an individual method to a substantial subsystem.)
  - Separate these requirements into individual clauses. For example:
    - Return value multiplied by itself must equal input AND
    - Must throw an exception if input is negative AND ....
  - Write a separate unit test for each clause, like the separate tests that Mort wrote in the story. That way, your set of tests is much more flexible, and easier to change when the requirements change.
- Work on the code in such a way as to satisfy a small number of these separate requirements at a time.
- Don't change or delete a unit test unless the corresponding requirement changes, or you find that the test does not correctly represent the intended requirement.

TESTING WITHIN A DEVELOPMENT TASK

The recommended cycle for a development task is therefore:

1. Check code out of source control.

2. Run the existing tests to make sure they pass. If you change the code and then find there are tests failing, you could spend a long time wondering what you did wrong.

3. Delete any existing unit tests for requirements that are no longer valid.
   For example, suppose Mort's requirement changes so that negative inputs just return a result of zero. He deletes the test ThrowsOnNegativeArgument, but keeps the BasicRooterTest.

4. **Loop:**
   {

   a. **Red: Write a new unit test and make sure it fails.**

      Write a new unit test:
      - To test the new feature that you're about to implement.
      - To extend the range of data that you use to test. For example, test a range of numbers rather than just one.
      - To exercise code that has not previously been exercised. See the section on code coverage below.

      Run the test and make sure that it fails. This is a good practice that avoids the mistake of forgetting to put an assertion at the end of the test method. If it definitely fails, then you know you've actually achieved something when you eventually get it to pass.

   b. **Green: Update your application to make the tests pass.**

      Make sure that all the tests pass—not just the new ones.

   c. **Refactor: Review the application code to make it easy to read and update.**

      Review the code to make sure that it's easy to read and update, and performs well. Then run the tests again.

   d. **Perform a code coverage check.**

   } **until** most of the code is covered by tests, and all the requirements are tested, and all the tests pass.

Code coverage

It is important to know how much of your code is exercised by the unit tests. Code coverage tools give you a measure of what percentage of your code has been covered by a unit test run and can highlight in red any statements that have not been covered.

Low coverage means that some of the logic of the code has not been tested. High coverage does not necessarily imply that all combinations of input data will be correctly processed by your code; but it nevertheless indicates that the likelihood of correct processing is good.

Aim for about 80%.

To see code coverage results, go to the *Unit Test* menu and choose **Analyze Code Coverage**. After you run tests, you'll see a table that shows the percentage of the code that has been covered, with a breakdown of the coverage in each assembly and method.

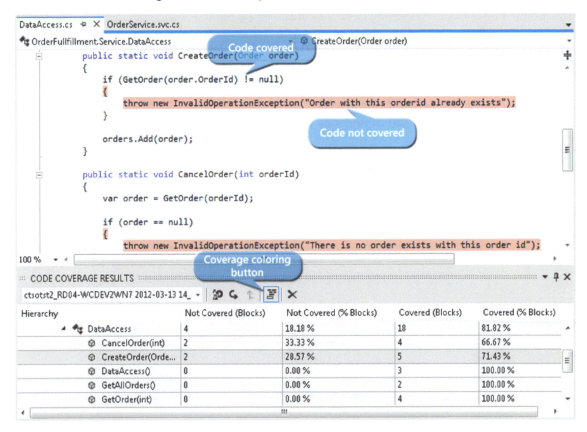

Code coverage results

Choose the **Coverage coloring** button to see the most useful feature, which shows you which bits of the code you have not exercised. Consider writing more tests that will use those parts.

Check-in policies

The best way to keep the server builds clean and green is to avoid checking bad code into source control. Set check-in policies to remind you and your team colleagues to perform certain tasks before checking in. For example, the testing policy requires that a given set of unit tests have passed. In addition to the built-in policies, you can define your own and download policies from the web. For more information, see the MSDN topic *Enhancing Code Quality with Team Project Check-in Policies*.

Users can override policies when they check in code; however, they have to write a note explaining why, and the event shows up in a report.

To set a check-in policy, on the **Team** menu in Visual Studio choose **Team Project Settings**, select **Source Control**. Click on the **Check-in Policy** tab.

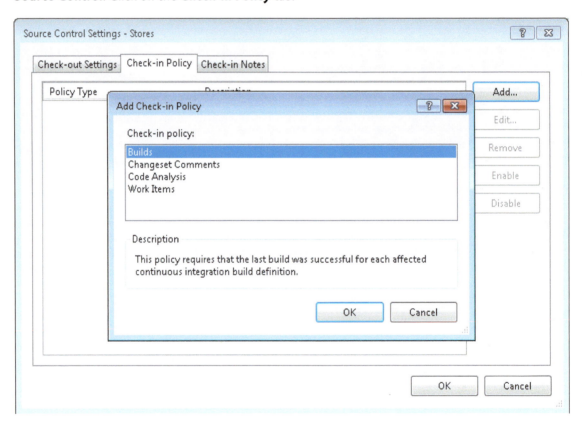

Add check-in policy

How to write good unit tests

A lot has been written about what makes a good unit test, and we don't have space to replicate it all here. If you're looking for more information, search the web for "unit test patterns."

However, there are some particularly useful tips.

Arrange – Act – Assert

The general form {Arrange; Act; Assert} is favored by many developers:

```
Arrange: Set up test data;
Act: Call the unit under test;
Assert: Compare the expected and actual results, and
log the result of the comparison as a fail or pass.
```

For example:

C#
```csharp
[TestMethod]
public void TestSortByFlavor()
{
 // Arrange: Set up test data:
    var catalog = new IceCreamCatalog(Flavor.Oatmeal, Flavor.Cheese);

 // Act: Exercise the unit under test:
    catalog.SortByFlavor();

 // Assert: Verify and log the result:
    Assert.AreEqual(Flavor.Cheese, catalog.Items[0].Flavor);
}
```

Test one thing with each test

Don't be tempted to make one test method exercise more than one aspect of the unit's behavior. Doing so leads to tests that are difficult to read and update. It can also lead to confusion when you are interpreting a failure.

Keep in mind that the MSTest test framework does not by default guarantee a specific ordering for the tests, so you cannot transfer state from one test to another. However, in Visual Studio, you can use the Ordered Test feature to impose a specific sequence.

Use test classes to verify different behavioral areas of the code under test

Separate tests for different behavioral features into different test classes. This usually works well because you need different shared utility methods to test different features. If you want to share some methods between test classes, you can of course derive them from a common abstract class.

Each test class can have a **TestInitialize** and **TestCleanup** method. (These are the MSTest attributes; there are equivalents in other test frameworks, typically named Setup and Teardown.) Use the initialize or setup method to perform the common tasks that are always required to set up the starting conditions for a unit test, such as creating an object to test, opening the database or connections, or loading data. The cleanup or teardown method is always called, even if a test method fails; this is a valuable feature that saves you having to write all your test code inside **try...finally** blocks.

TEST EXCEPTION HANDLING

Test that the correct exceptions are thrown for invalid actions or inputs.

You could use the **[ExpectedException]** attribute, but be aware that a test with that attribute will pass no matter what statement in the test raises an exception.

A more reliable way to test for exceptions is shown here:

```csharp
C#
    [TestMethod, Timeout(2000)]
    public void TestMethod1()
    {   ...
        AssertThrows<InvalidOperationException>( delegate
         {
            MethodUnderTest();
         });
    }


    internal static void AssertThrows<exception>(Action method)
                            where exception : Exception
    {
        try
        {
            method.Invoke();
        }
        catch (exception)
        {
            return; // Expected exception.
        }
        catch (Exception ex)
        {
            Assert.Fail("Wrong exception thrown: " + ex.Message);
        }
        Assert.Fail("No exception thrown");
    }
```

A function similar to AssertThrows is built into many testing frameworks.

DON'T ONLY TEST ONE INPUT VALUE OR STATE

By verifying that 2.0==MySquareRootFunction(4.0), you haven't truly verified that the function works for all values. The code coverage tool might show that all your code has been exercised, but it might still be the case that other inputs, or other starting states, or other sequences of inputs, give the wrong results.

Therefore, you should test a representative set of inputs, starting states, and sequences of action.

Look for boundary cases: those where there are special values or special relationships between the values. Test the boundary cases, and test representative values between the boundaries. For example, inputs of 0 and 1 might be considered boundary cases for a square root function, because there the input and output values are equal. So test, for example, -10, -1, -0.5, 0, 0.5, 1, and 10.

Test also across the range. If your function should work for inputs up to 4096, try 4095 and 4097.

The science of *model-driven testing* divides the space of inputs and states by these boundaries, and seeks to generate test data accordingly.

For objects more complex than a simple numeric function, you need to consider relationships between different states and values: for example, between a list and an index of the list.

Separate test data generation from verification

A postcondition is a Boolean expression that should always be true of the input and output values, or the starting and ending states of a method under test. To test a simple function, you could write:

```C#
[TestMethod]
public void TestValueRange()
{
  while (GenerateDataForThisMethod(
            out string startState, out double inputValue)))
  {
     TestOneValue(startState, inputValue);
  }
}
// Parameterized test method:
public void TestOneValue(string startState, double inputValue)
{
    // Arrange - Set up the initial state:
    objectUnderTest.SetKey(startState);

    // Act - Exercise the method under test:
    var outputValue = objectUnderTest.MethodUnderTest(inputValue);

    // Assert - Verify the outcome:
    Assert.IsTrue(PostConditionForThisMethod(inputValue, outputValue));
}

// Verify the relationship between input and output values and states:
private bool
    PostConditionForThisMethod
        (string startState, double inputValue, double outputValue)
{
    // Accept any of a range of results within specific constraints:
    return startState.Length>0 && startState[0] == '+'
                ? outputValue > inputValue
                : inputValue < outputValue;
}
```

To test an object that has internal state, the equivalent test would set up a starting state from the test data, and call the method under test, and then invoke the postcondition to compare the starting and ending states.

The advantages of this separation are:

- The postcondition directly represents the actual requirement, and can be considered separately from issues of what data points to test.
- The postcondition can accept a range of values; you don't have to specify a single right answer for each input.
- The test data generator can be adjusted separately from the postcondition. The most important requirement on the data generator is that it should generate inputs (or states) that are distributed around the boundary values.

Pex generates test data

Take a look at *Pex*, which is an add-in for Visual Studio.

There's also a *standalone online version*.

Pex automatically generates test data that provides high code coverage. By inspecting the code under test, Pex generates interesting input-output values. You provide it with the parameterized version of the test method, and it generates test methods that invoke the test with different values.

(The website also talks about Moles, an add-in for Visual Studio 2010, which has been replaced by an integrated feature, fakes, for Visual Studio 2012.)

Isolation testing: fake implementations

If the component that you are developing depends on another component that someone else is developing at the same time, then you have the problem that you can't run your component until theirs is working.

A less serious issue is that even if the other component exists, its behavior can be variable depending on its internal state, which might depend on many other things; for example, a stock market price feed varies from one minute to the next. This makes it difficult to test your component for predictable results.

The solution is to isolate your component by replacing the dependency with a *fake* implementation. A fake simulates just enough of the real behavior to enable tests of your component to work. The terms *stub*, *mock*, and *shim* are sometimes used for particular kinds of fakes.

The principle is that you define an interface for the dependency. You write your component so that you pass it an instance of the interface at creation time. For example:

```C#
// This interface enables isolation from the stock feed:
public interface IStockFeed
{
  int GetSharePrice(string company);
}
```

```
// This is the unit under test:
public class StockAnalyzer
{
    private IStockFeed stockFeed;

    // Constructor takes a stockfeed:
    public StockAnalyzer(IStockFeed feed) { stockFeed = feed; }

    // Some methods that use the stock feed:
    public int GetContosoPrice() { ... stockFeed.GetSharePrice(...) ... }
}
```

By writing the component in this way, you make it possible to set it up with a fake implementation of the stock feed during testing, and a real implementation in the finished application. The key thing is that the fake and the real implementation both conform to the same interface. If you like interface diagrams, here you go:

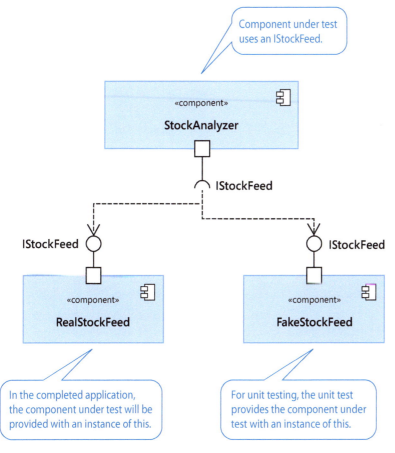

Interface injection

This separation of one component from another is called "interface injection." It has the benefit of making your code more flexible by reducing the dependency of one component on another.

You could define **FakeStockFeed** as a class in the ordinary way:

```C#
// In test project.
class FakeStockFeed : IStockFeed
{
    public int GetSharePrice (string company) { return 1234; }
}
```

And then in your test, you'd set up your component with an instance of the fake:

```C#
    [TestClass]
    public class StockAnalysisTests
    {
        [TestMethod]
        public void ContosoPriceTest()
        {
            // Arrange:
            var componentUnderTest = new StockAnalyzer(new FakeStockFeed());
            // Act:
            int actualResult = componentUnderTest.GetContosoPrice();
            // Assert:
            Assert.AreEqual(1234, actualResult);
        }
    }
```

However, there's a neat mechanism called Microsoft Fakes that makes it easier to set up a fake, and reduces the clutter of the fake code.

MICROSOFT FAKES

If you're using MSTest in Visual Studio 2012, you can have the stub classes generated for you.

In Solution Explorer, expand the test project's references, and select the assembly for which you want to create stubs—in this example, the Stock Feed. You can select another project in your solution, or any referenced assembly, including system assemblies. On the shortcut menu, choose **Add Fakes Assembly**. Then rebuild the solution.

Now you can write a test like this:

```C#
[TestClass]
class TestStockAnalyzer
{
    [TestMethod]
    public void TestContosoStockPrice()
    {
      // Arrange:

        // Create the fake stockFeed:
        IStockFeed stockFeed =
            new StockAnalysis.Fakes.StubIStockFeed() // Generated by Fakes.
                {
                    // Define each method:
                    // Name is original name + parameter types:
                    GetSharePriceString = (company) => { return 1234; }
                };

        // In the completed application, stockFeed would be a real one:
        var componentUnderTest = new StockAnalyzer(stockFeed);

      // Act:
        int actualValue = componentUnderTest.GetContosoPrice();

      // Assert:
        Assert.AreEqual(1234, actualValue);
    }
    ...
}
```

The special piece of magic here is the class **StubIStockFeed**. For every public type in the referenced assembly, the Microsoft Fakes mechanism generates a stub class. The type name is the same as the original type, with "Stub" as a prefix.

This generated class contains a delegate for each message defined in the interface. The delegate name is composed of the name of the method plus the names of the parameter types. Because it's a delegate, you can define the method inline. This avoids explicitly writing out the class in full.

Stubs are also generated for the getters and setters of properties, for events, and for generic methods. Unfortunately IntelliSense doesn't support you when you're typing the name of a delegate, so you will have to open the Fakes assembly in Object Browser in order to check the names.

You'll find that a .fakes file has been added to your project. You can edit it to specify the types for which you want to generate stubs. For more details, see *Isolating Unit Test Methods with Microsoft Fakes*.

Mocks

A mock is a fake with state. Instead of giving a fixed response to each method call, a mock can vary its responses under the control of the unit tests. It can also log the calls made by the component under test. For example:

```C#
[TestClass]
class TestMyComponent
{

    [TestMethod]
    public void TestVariableContosoPrice()
    {
     // Arrange:
        int priceToReturn;
        string companyCodeUsed;
        var componentUnderTest = new StockAnalyzer(new StubIStockFeed()
            {
                GetSharePriceString = (company) =>
                    {
                        // Log the parameter value:
                        companyCodeUsed = company;
                        // Return the value prescribed by this test:
                        return priceToReturn;
                    };
            };
        priceToReturn = 345;
     // Act:
        int actualResult = componentUnderTest.GetContosoPrice(priceToReturn);
     // Assert:
        Assert.AreEqual(priceToReturn, actualResult);
        Assert.AreEqual("CTSO", companyCodeUsed);
    }
...}
```

Shims

Stubs work if you are able to design the code so that you can call it through an interface. This isn't always practical, especially when you are calling a platform method whose source you can't change.

For example, **DateTime.Now** is not accessible for us to modify. We'd like to fake it for test purposes, because the real one inconveniently returns a different value at every call. So we'll use a shim:

```csharp
C#
[TestClass]
public class TestClass1
{
        [TestMethod]
        public void TestCurrentYear()
        {
            using (ShimsContext.Create())
            {
              // Arrange:
                // Redirect DateTime.Now to return a fixed date:

                System.Fakes.ShimDateTime.NowGet = () =>
                { return new DateTime(2000, 1, 1); };

                var componentUnderTest = new MyComponent();

              // Act:
                int year = componentUnderTest.GetTheCurrentYear();

              // Assert:
                Assert.AreEqual(2000, year);
            }
        }
}

// Code under test:
public class MyComponent
{
    public int GetTheCurrentYear()
    {
      // During testing, this call will be redirected to the shim:
      DateTime now = DateTime.Now;
      return now.Year;
    }
}
```

What happens is that any call to the original method gets intercepted and redirected to the shim code.

Shims are set up in the same way as stubs. For example, to create a shim for **DateTime**, begin by selecting the reference to System in your test project, and choose **Add Fakes Assembly**.

Notice the ShimsContext: when it is disposed, any shims you created while it was active are removed.

Shim class names are made up by prefixing "Shim" to the original type name.

You might see an error stating that the Fakes namespace does not exist. Fix any other errors, and this will then work correctly.

The shim in this example modifies a static property. You can also create shims to intercept calls to constructors; and to methods or properties of all instances of a class, or to specific instances. See the MSDN topic *Using shims to isolate calls to non-virtual functions in unit test methods*.

Shims in Visual Studio 2010

In Visual Studio 2010, you have to obtain an add-in called Moles. The naming conventions are slightly different: the stub method has "M" instead of "Shim" as its prefix. Also, you must apply an attribute to the test class:

```C#
[HostType("Moles")]
[TestClass]
public class TestClass1
{
    [TestInitialize]
    public void TestSetup()
    {
        // Redirect DateTime.Now to return a fixed date:

        MDateTime.NowGet = () => { return new DateTime(2000, 1, 1); } ;
    }
}
```

The Moles add-in injects patches into the code under test, intercepting calls to the methods that you specify. It generates a set of mirrors of the classes that are referenced by the code under test. To specify a particular target class, prefix its name with "M" like this: MDateTime or System.MFile. You can target classes in any assembly that is referenced from yours. Use IntelliSense to choose the method that you want to redirect. Property redirects have "Get" and "Set" appended to the property's name. Method redirects include the parameter type names; for example MDateTime.IsLeapYearInt32.

Just as with mocks, you can get your mole to log calls and control its response dynamically from the test code.

> **Note:** *Moles in Visual Studio 2010 need rework to turn them into fakes for Visual Studio 2012.*

Testing and debugging

When you run unit tests normally, we recommend that you use the command that runs without the debugger. Typically you expect the tests to pass, and debugging just slows things down.

If a test fails, if you are using the Visual Studio built-in MSTest framework, use the **Debug Checked Tests** command in the Test Results view; this will rerun the tests that failed.

If you are using another test framework, you can usually find add-ins that integrate the test framework with Visual Studio debugging.

Without such an add-in, to run the debugger with another testing framework, the default steps are:

1. Set the test project as the solution's Startup project
2. In the test project's properties, on the **Debug** tab, set the **Start Action** to start your test runner, such as NUnit, and open it on the test project.
3. When you want to run tests, start debugging from Visual Studio, and then select the tests in the test runner.

IntelliTrace

IntelliTrace keeps a log of the key events as your tests and code execute in debug mode, and also logs variable values at those events. You can step back through the history of execution before the test failed, inspecting the values that were logged.

To enable it, on the Visual Studio **Debug** menu, choose **Options and Settings, IntelliTrace**. You can also vary the settings to record more or less data.

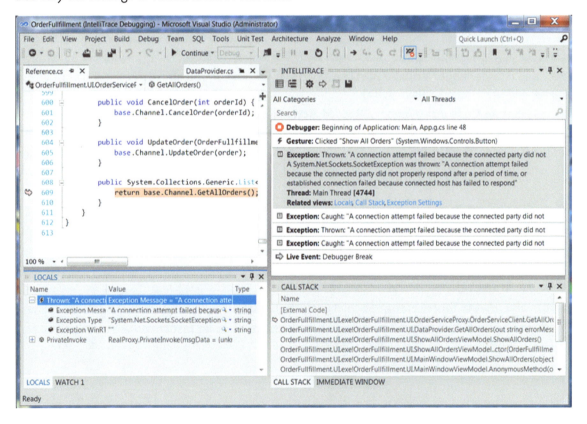

Using IntelliTrace

IntelliTrace is particularly useful when the developer debugging the code was not present when the bug was discovered. When we come to discuss the build service, we will see how a test failure can automatically save the state of the environment at the failure, and attach the state to a bug work item. During manual testing, the tester can do the same at the touch of a button. IntelliTrace enables the developer to look not just at the state of the system at the instant of failure, but also at the states that led up to that point.

Coded UI tests

Unit tests typically work by calling methods in the interface of the code under test. However, if you have developed a user interface, a complete test must include pressing the buttons and verifying that the appropriate windows and content appear. Coded UI tests (CUITs) are automated tests that exercise the user interface. See the MSDN topic *Testing the User Interface with Automated Coded UI Tests*.

HOW TO CREATE AND USE CODED UI TESTS

Create a coded UI test

To create a coded UI test, you have to create a Coded UI Test Project. In the **New Project** dialog, you'll find it under either **Visual Basic\Test** or **Visual C#\Test**. If you already have a Coded UI Test project, add to it a new Coded UI Test.

In the **Generate Code** dialog, choose **Record Actions**. Visual Studio is minimized and the Coded UI Test builder appears at the bottom right of your screen.

Choose the **Record** button, and start the application you want to test.

Recording a coded UI test

Perform a series of actions that you want to test. You can edit them later.

You can also use the **Target** button to create assertions about the states of the UI elements.

The **Generate Code** button turns your sequence of actions into unit test code. This is where you can edit the sequence as much as you like. For example, you can delete anything you did accidentally.

Running coded UI tests

Coded UI tests run along with your other unit tests in exactly the same way. When you check in your source code, you should check in coded UI tests along with other unit tests, and they will run as part of your build verification tests.

> **Tip:** *Keep your fingers off the keyboard and mouse while a CUIT is playing. Sitting on your hands helps.*

Edit and add assertions

Your actions have been turned into a series of statements. When you run this test, your actions will be replayed in simulation.

What's missing at this stage is assertions. But you can now add code to test the states of UI elements. You can use the **Target** button to create proxy objects that represent UI elements that you choose. Then you write code that uses the public methods of those objects to test the element's state.

Extend the basic procedure to use multiple values

You can edit the code so that the procedure you recorded will run repeatedly with different input values.

In the simplest case, you simply edit the code to insert a loop, and write a series of values into the code.

But you can also link the test to a separate table of values, which you can supply in a spreadsheet, XML file, or database. In a spreadsheet, for example, you provide a table in which each row is a set of data for each iteration of the loop. In each column, you provide values for a particular variable. The first row is a header in which the data names are identified:

Flavor	Size
Oatmeal	Small
Herring	Large

In the Properties of the coded UI test, create a new **Data Connection String**. The connection string wizard lets you choose your source of data. Within the code, you can then write statements such as

C#
```
var flavor = TestContext.DataRow["Flavor"].ToString();
```

Isolate

As with any unit tests, you can isolate the component or layer that you are testing—in this case, the user interface—by providing a fake business layer. This layer should simply log the calls and be able to change states so that your assertions can verify that the user interface passed the correct calls and displayed the state correctly.

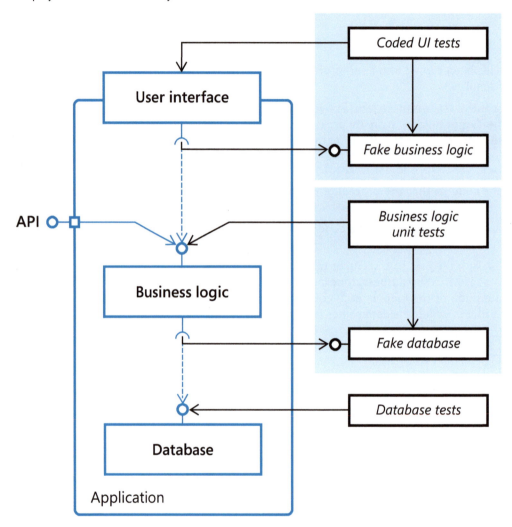

Well-isolated unit tests

Test first?

You might think this isn't one of those cases where you can realistically write the tests before you write the code. After all, you have to create the user interface before you can record actions in the Coded UI Test Builder.

This is true to a certain extent, especially if the user interface responds quite dynamically to the state of the business logic. But nevertheless, you'll often find that you can record some actions on buttons that don't do much during your recording, and then write some assertions that will only work when the business logic is coupled up.

Coded UI tests: are they unit or system tests?

Coded UI tests are a very effective way of quickly writing a test. Strictly speaking, they are intended for two purposes: testing the UI by itself in isolation (with the business logic faked); and system testing your whole application (which we'll discuss in Chapter 5, "Automating System Tests").

But coded UI tests are such a fast way of creating tests that it's tempting to stretch their scope a bit. For example, suppose you're writing a little desktop application—maybe it accesses a database or the web. The business logic is driven directly from the user interface. Clearly, a quick way of creating tests for the business logic is to record coded UI tests for all the main features, while faking out external sources of variation such as the web or the database. And you might decide that your time is better spent doing that than writing the code for the business logic.

Cover your ears for a moment against the screams of the methodology consultants. What's agonizing them is that if you were to test the business logic by clicking the buttons of the UI, you would be coupling the UI to the business logic and undoing all the good software engineering that kept them separate. If you were to change your UI, they argue, you would lose the unit tests of your business logic.

Furthermore, since coded UI tests can only realistically be created after the application is running, following this approach wouldn't allow you to follow the test-first strategy, which is very good for focusing your ideas and discussions about what the code should do.

For these reasons, we don't really recommend using coded UI tests as a substitute for proper unit tests of the business logic. We recommend thinking of the business logic as being driven by an API (that you could drive from another code component), and the UI as just one way of calling the operations of the API. And to write an API, it's a good idea to start by writing samples of calling sequences, which become some of your test methods.

But it's your call; if you're confident that your app is short-lived, small, and insignificant, then coded UI tests can be a great way to write some quick tests.

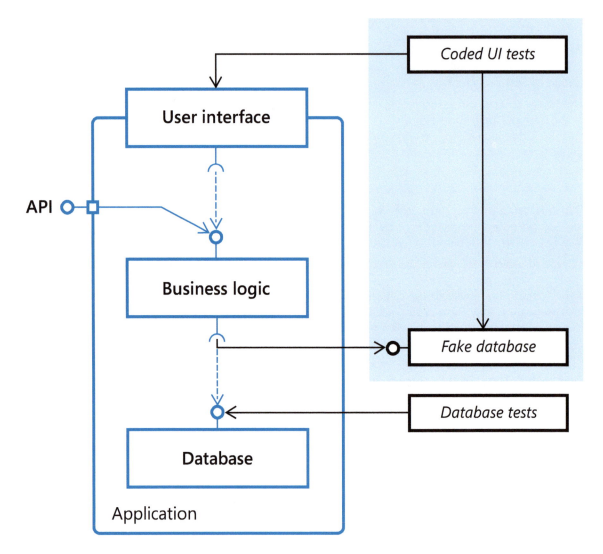

Coded UI test for application testing

Unit Testing: Testing the Inside

Designing for coded UI tests

When you run a test, the CUIT engine has to find each control that your actions use. It does so by navigating the presentation tree, using the names of the UI elements. If the user interface is redesigned, the tests might not work because the elements cannot be found. Although the engine has some heuristics for finding moved elements, you can improve its chances of working.

- In HTML, make sure every element has an ID.
- In Windows presentation technologies, support Accessibility.
- If you design a custom control, define a CUIT extension to help the recorder interpret user gestures (see the MSDN topic *Enable Coded UI Testing of Your Custom Controls*). For example, when you use a file selection control, the recorder does not record a sequence of mouse clicks, but instead records which file was selected. In the same way, you can define a recorded extension that encodes the user's intentions when using your control.

Maintaining coded UI tests

A drawback of CUITs is that they must be recreated whenever there are significant changes to the user interface definition. You can minimize the effort needed:

- Make separate recordings, and thereby separate methods, for different forms or pages, and for groups of no more than a dozen actions.
- If a change occurs, locate the affected methods and rerecord just those methods.
- Use the CUIT Editor to update the code. It is also possible to edit the code directly, but the result is more reliable using the editor.

This is a brief overview of CUITs. For more information, see the MSDN topic *How to: Edit a Coded UI Test Using the Coded UI Test Editor*.

Continuous integration with build verification tests

Build verification tests are sometimes called the rolling build or the nightly build. On a regular or a continuous basis, the build service compiles and tests the software that has been checked into the source tree. If a test fails—or worse, if the source doesn't compile—then the service sends plaintive emails to everyone.

Source control helps team members avoid overwriting each other's work, and lets a team of people work on a single body of software. As soon as you have installed TFS and created a team project, you can use Visual Studio to create a source tree and add code to it, and assign permissions to other users.

But to make sure that the code does what is expected of it, you must set up regular builds. Typically, you will have more than one set up: a continuous ("rolling") build that runs most of the unit tests; and a nightly build that runs more extensive tests, including performance and load tests, and automated systems tests (which we will discuss in the next chapter).

If you're a test professional, you won't need us to tell you this, but just to confirm the point: The only way to deliver quality software on time is to never let code be checked in without good test coverage; to run the tests all the time during development; and never to let bugs go unfixed. Anyone who has been around a while knows of projects where they let a moraine of untested code be pushed back towards the end of the project, and knows the pain that caused.

So these are the rules about making changes to the source:

- Check in all your code at least every week, and preferably more often. Plan your development tasks so that you can contribute a small but complete extension or improvement to the system at each check-in.

- Before checking in a change, use the Get Latest Version command (on the shortcut menu of your project in Solution Explorer) to integrate updates that have been made by other team members while you were working on your changes. Rebuild everything and run the unit tests.

- Use the **Run All Impacted Tests** command, which determines what tests are affected by changes that you have made or imported. (See *Streamline Testing Process with Test Impact Analysis* on MSDN.) Your changes can affect tests that you didn't write, and changes made by others can affect tests you have written.
  To use this command, you must initialize a baseline when you check out code. (See the MSDN topic *How to: Identify the Test Impact of Code Changes During Development*.)

- Switch on **Code Coverage** and check that at least 80% of your code has been exercised by the tests. If not, use the coloring feature to find code that has not been used. Write more tests.

- Do not check in your changes unless 80% coverage has been achieved and the tests all pass. Some instances of lower coverage are allowed. For example, where code is generated, it is sometimes reasonable to take coverage of one generated item as verification of another. But if you propose to make an exception, the proposal must be reviewed by a colleague with a skeptical personality.
  To enforce this rule, create a testing check-in policy.

- If the rolling build breaks after you checked in code, you (and everyone else) will get a notification by email. If your changes might be the cause of the problem, undo your check-in. Before working on any other code, check in a fixed version. If it turns out that the fix will take some time, reopen the work item related to this change; you can no longer claim it is complete.

- Don't go home immediately after checking in. You don't want to come back in the morning to find everyone's mailbox full of build failures.
  This doesn't apply if your team uses gated check-ins, where your code isn't actually integrated into the main body of the source until the tests pass on an auxiliary build server.

How to set up a build in Team Foundation Server

In Team Explorer, in the **Builds** window, choose **New Build Definition**.

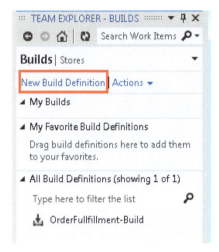

New build definition

In the wizard, you can select how you want it to start. Typically, you'd choose a Scheduled build to run the biggest set of tests every night. Continuous integration runs a build for every check-in; rolling builds require fewer resources, running no more than one at a time. Gated builds are a special case: each check-in is built and tested on its own, and is only merged into the source tree if everything passes.

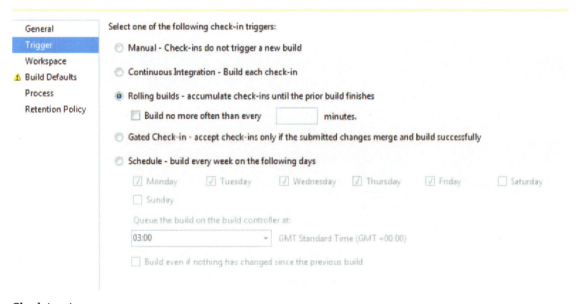

Check-in triggers

Create a drop folder on a suitable machine, and set sharing permissions so that anyone on your project can read it and the build service can write to it.

Specify this folder in the build definition:

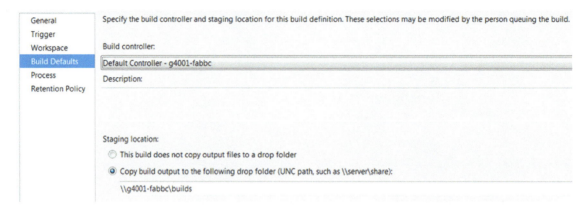

Create a drop folder

Under **Process**, you can leave the default settings, though you might want to check them.

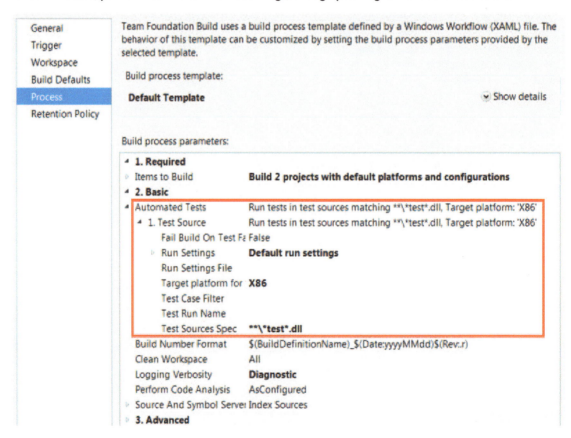

Automated test settings

Save the definition.

The build definition appears in Team Explorer. You can run any build on demand.

Code coverage

You'll want to make sure that code coverage analysis is included in the project settings so that you can see code coverage results in the build reports. Enable code analysis in your project's properties.

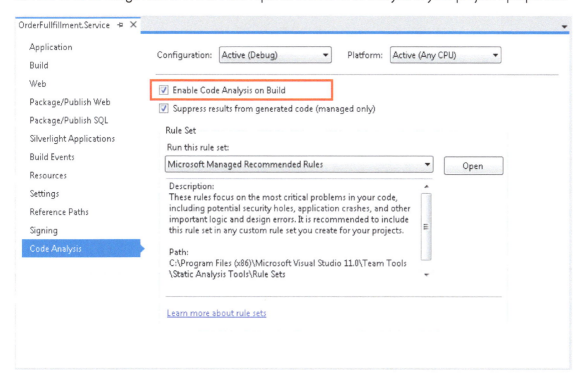

Code coverage analysis

Third-party build frameworks

You can get extensions that allow you to execute the build with third party frameworks. For example, if you use JUnit, or if you perform builds using Ant or Maven, you can integrate them with Team Foundation Server. Go to the MSDN Visual Studio Gallery and search for the page: *Team Foundation Server Build Extensions Power Tool* for more information.

Generate an installer from the build

Each build should generate an installer of the appropriate type—typically a Microsoft Windows Installer (setup) package, but it could also be, for example, a Visual Studio Extension (.vsix) or, for websites, a Microsoft Internet Information Services (IIS) deployment package. The type of installer you need varies with the type of project.

For a website, choose **Publish** from the shortcut menu of your Visual Studio project. The Publish Web Application wizard will walk you through generating the installer. The installer will be generated in the output folder of the project. The regular build on the build server will create an installer in the same way.

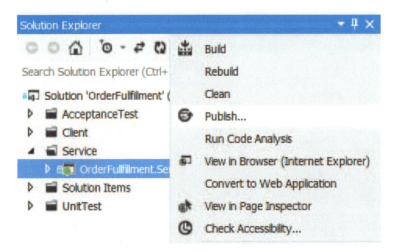

Creating a website installer

For a desktop application, you have to add an installer project to your solution. In the **New Project** dialog, look under **Other Project Types, Setup**. The details are in the MSDN page *Windows Installer Deployment*. Set the project properties to define what you want to include in the installer. When the project is built—either on your development machine or in the build server—the output folder of that project will contain a setup.exe and associated folders, which can be copied to wherever you want to deploy.

For a distributed system, you will need more than one setup project, and the automated build will need a deployment script. We'll discuss that in a later chapter.

In addition, there is a separate publication mechanism, ClickOnce Deployment. We'll discuss this more in Chapter 5, "Automating System Tests." There is additional information in the MSDN topic *Click-Once Security and Deployment*.

TEST THE INSTALLER

Write a basic unit test to verify that an installer file was successfully generated. (The main unit tests aren't, of course, dependent on the installer; they work by directly running the .dll or .exe that was created by the compiler.)

Why do we recommend you generate an installer from every build?

Firstly, there are many bugs that emerge when you deploy your system in an environment that is not the build machine. In the next chapter, we're going to recommend that system testing should always begin by installing the system on a clean machine.

Secondly, in good iterative software development practice, you should deliver something to your stakeholders at the end of each iteration. The installer is what you're delivering. When they install and run it, they should be able to work through the user stories that were planned at the start of the iteration.

Maybe we should add, for those with long memories, that generating installers is a lot easier in Visual Studio 2012 than it used to be back in the bad old days.

Monitoring the build

There's no point in having a regular build unless you know whether it passed or failed. There are four ways to find out:

- **Build Notifications Tool** on your desktop. Once you've pointed it to your server, it sits hidden in your taskbar and pops up toast when a build completes. You can set it just to show builds initiated by you or your check-ins.
  After you check in some code, it's reassuring, a while later, to see the "build OK" flag appear in the corner of your screen.
  You'll find it on the **Start** menu under **Microsoft Visual Studio** > **Team Foundation Server Tools**. Before you can point it to a particular server, you must have connected to that server at least once using Team Explorer.

- **Email**. You can get notifications sent to the team. This is particularly useful if a build fails, because you want to fix it as a matter of high priority. Unlike the build notification tool, you won't miss it if you're away from your desktop for a while.
  To set up email notifications, see the next section.

- **Build Explorer**. Monitor the build by opening the **Builds** node in Team Explorer. You can also define a new build and start a build run from here.

- **Build Reports** appear under the **Reports** node in Team Explorer. A particularly useful report is Build Quality Indicators. It tells you about test coverage in recent builds. Make sure the coverage remains high.
  If you can't see the Reports node, look again at the section about enabling reports in the previous chapter.

How to set up build failure or completion emails

You need to enable emails on the server, and then set up the details from a client machine.

On the Team Foundation Server machine

Open Team Foundation Server Administration console, and under your server machine name, select **Application Tier**.

Make sure that **Service Account** is set to an account that has permission on your domain to send email messages. If it isn't, use **Change Account**. Don't use your own account name, because you will change your password from time to time. (If you ever have to change the password on the service account, notice that there's an Update Password command. This doesn't change the password on that account; it changes the password that Team Foundation Server presents when it tries to use that account.)

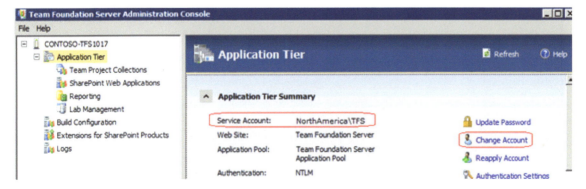

Service account settings

Scroll down to **Email Alert** Settings and choose **Alert Settings**. At this point, you'll need the address of an SMTP server in your company's network. Get one from your domain administrator.

Email alert settings

On your desktop machine

From Team Explorer, choose **Home, Settings, Project Alerts**. Your team project site will open in your browser on your own alerts. To set the team alerts choose the **Administer Server** icon at the top right or the **Advanced Alerts Management Page** link.

Administer server

Choose **alerts, My Alerts, Build Alerts**, and then **New**. In the dialog box, select an alert template to create an alert for **A build fails**. Add a new clause to the query: **Requested For | Contains | [Me]**.

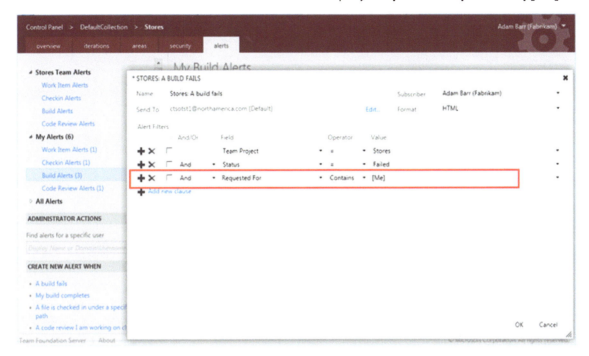

Build alerts

To set up build alerts in Visual Studio 2010

Install the Team Foundation Server Power Tools. On your development machine, go to the *Visual Studio Gallery* and find *Team Foundation Server Power Tools*. Close Visual Studio before installing.

Open Visual Studio and from the **Team** menu open **Alerts Explorer**. In the toolbar of the Alerts Explorer window, choose **New Alert**. Under **Build Alerts**, choose **Failed Build**. In the next dialog, provide your team's group email address.

Responding to build failure alarms

When you get an email notification, click the link near the top of the message:

Subject:	Contoso-ctso3-p1 Build Deadline rolling_20111129.3 partially succeeded

Contoso-ctso3-p1 Build Deadline rolling_20111129.3 partially succeeded

Team Project:	Contoso-ctso3-p1
Build Number:	Deadline rolling_20111129.3
Build Controller:	Default Controller - contoso-tfs1017
Build Definition:	\Contoso-ctso3-p1\Deadline rolling
Build started by:	
Build Start Time:	11/29/2011 11:32:58 AM
Build Finish Time:	11/29/2011 11:33:45 AM
Build Log Location:	\\contoso-tfs1017\drops\Deadline rolling\Deadline rolling_20111129.3\logs\Deadline.log

Email notification

In the build summary, you'll see **Associated Changesets**. This is the list of changes that were made since the previous successful build. One of them almost certainly caused the failure. The person responsible for each changeset is also listed. Enquire of them what they propose to do.

If it's you, undo your recent change, fix the problem, and then check in again.

The builds check-in policy

You might be tempted to fix the problem and check in the fix without first undoing the change. Maybe you like working under pressure with alarms going off every few minutes as someone checks in another change. But it's not an advisable tactic, because while your bug is in the build, it masks any other problem. And of course a quick fix often turns out not to be as quick as you first thought, and can be unreliable. So if the build on the server fails, it's better to back out the associated changesets that are listed in the log, to get the build back to green. Then you can think about the issue calmly.

You can set a check-in policy that insists on this procedure. In **Visual Studio** on the **Team** menu, choose **Team Project Settings**, **Source Control**. (If you can't see that menu item, make sure that Team Explorer is connected to your project.) In the dialog, in the **Check-in Policies** tab, add the **Builds** policy. Then if anyone tries to check in code while a build is failing that uses that code, they get a discouraging message. Like other check-in policies, you can override it if you must. But the idea is that while the build is failing, you should fix it only by undoing recent changes.

The build report

You can obtain the results of the latest server builds through the Builds section of Team Explorer. The project portal website also includes reports of the build results. (See the MSDN topic *Build Dashboard (Agile)*.)

In addition to the automatically generated test pass/fail results, the build report also shows a result that you can set manually in the log of each run. Typically, a member of the test or development team will regularly "scout" for a recent build that completed properly, that has good performance, and is suitable for further testing.

When you want to deploy your system for manual testing or for creating samples or demonstrations, look at the report of recent builds. The report includes links to the folder where you can find the built assemblies or installers.

Spreading the load of many unit tests

If you have more than a few hundred unit tests, you can spread the load of running them across multiple machines. (See the MSDN topic, *Running Unit Tests on Multiple Machines Using a Test Controller and Test Agents*.)

To spread the load, set up several machines that contain test agents, and a machine that has a test controller. These items can be found on the Visual Studio installation DVD. We will meet them again when we discuss setting up a lab environment. Configure the test agents to talk to the test controller. In Visual Studio, in the **Test** menu, open **Manage Test Controller** and select your controller.

When you run the unit tests, they will run on the machines that contain the test agents. The test controller will separate the tests into batches of 100, and run each batch on a different machine.

Summary

Unit testing provides a crucial engineering practice, ensuring not only that the system being built compiles correctly, but also that the expected behavior can be validated in every build. Visual Studio 2012 provides the following capabilities to an application development lifecycle:

- Unit testing is integrated into the development environment.
- Code coverage reports help you test every code path.
- Fakes allow you to isolate units, allowing parallel development of units.
- Coded UI Tests create test code from recorded manual test sessions.
- Integration with third-party unit testing frameworks.
- IntelliTrace reports help you find out how a fault occurred.
- Continuous integration build service.
- Build reports and alerts show you anything that fails.
- Automatic load spreading when there are many unit tests.

Differences between Visual Studio 2010 and Visual Studio 2012

In Visual Studio 2012:

- Unit Test Runner. The interface for running unit tests is substantially changed from Visual Studio 2010. It isn't too difficult to find your way around either of them. However, you can write test methods using the MSTest framework (where tests are introduced with the **[TestMethod]** attribute) in exactly the same way. For more information, see the MSDN topic *Running Unit Tests with Test Explorer*.

- Third-party test frameworks such as NUnit are supported. The tests from any framework appear in Unit Test Explorer, provided there is an adapter for it. Adapters for several popular frameworks are available, and you can write your own. In Visual Studio 2010, tests in other frameworks have to be run from their own user interface, although you can get add-ins to improve the integration. See the MSDN topic, How to: *Install Third-Party Unit Test Frameworks*.

- Fakes (stubs and shims) are built-in features. For Visual Studio 2010, you have to get the Moles add-in, which is not compatible with Fakes. See the MSDN topic, *Isolating Unit Test Methods with Microsoft Fakes*.

- C++ and native code tests can be created and run. They are not available in Visual Studio 2010. See *Writing Unit tests for C/C++ with the Microsoft Unit Testing Framework for C++* on MSDN.

- Unit test projects. There are separate project types for unit tests, coded UI tests, load tests, and so on. In Visual Studio 2010, there is just one type into which you can put different types of test. See the MSDN topic, *How to: Create a Unit Test Project*.

- **Windows apps** are supported with specialized unit testing features.

- Compatibility. Unit tests and test projects created in Visual Studio 2010 will run on Visual Studio 2012. You can't use the Visual Studio 2010 Express edition for unit tests. See *Upgrading Unit Tests from Visual Studio 2010 on MSDN*.

Where to go for more information

All links in this book are accessible from the book's online bibliography available on MSDN: *http://msdn.microsoft.com/en-us/library/jj159339.aspx.*

3   Lab Environments

Testing can be a fascinating and exciting job (tell your kids), but like any occupation it has its tedious moments. Setting up hardware and installing your software product so that you can test it are not the exciting parts. In Visual Studio, *lab environments* take a lot of the tedium out of setting up fresh computers and configuring them for testing.

In Chapter 2, "Unit Testing: Testing the Inside," we looked at unit tests and at how to run them on the build service. As we noted, unit tests are great for guarding against regressions if you run a lifecycle in which you often revise parts of the application code. They are less effective for finding new bugs; manual testing is better for that purpose.

Unit tests have another drawback: usually they run on a single machine. But if you want to run under more realistic conditions, your testing should begin by installing the product in the same way that the users will install it. You should run the product in an environment similar to the operational one. If the product is a distributed application, you should install the different components on different computers.

At Contoso, when the team wants to begin system testing, an IT professional is assigned to find and allocate a collection of machines on which the system can be run. If it's a distributed system, there might be wiring to do. The Contoso corporate network is averse to test machines, so a private network has to be configured, and that includes a domain name server.

This setup is very time-consuming and error-prone. In addition, if a product is to be tested again after an interval of a few months, it might be difficult to replicate the original environment exactly—the person who tested it first might have moved on, or lost their notes. Some tests might therefore yield results that are difficult to compare to the originals. All this makes it tough for Contoso to operate on a rapid cycle; setting up an environment to test a new feature or a bug fix might easily cost a lot more, both in time and in other resources, than updating the application code.

Fortunately their colleagues from Fabrikam can show Contoso a better way. Visual Studio Lab Management performs not only the mundane task of assigning machines to test environments, but also helps you to create new environments, configure them for tests, deploy your software, and run tests.

A new lab environment consisting of several machines can be ready within minutes. This makes a considerable difference to the total time taken to run tests.

What is a lab environment?

A lab environment is a collection of computers that are managed as a single unit, and on which you deploy the system under test along with test software. Here is a typical configuration of machines in a lab environment:

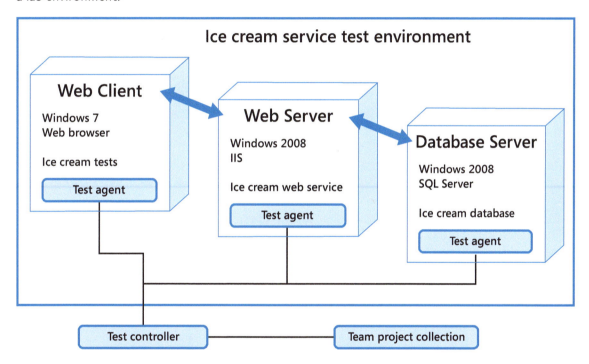

Typical lab environment configuration

This one is set up for automated tests of an ice cream vending service. The software product itself consists of a web service that runs on Internet Information Services (IIS) and a database that runs on a separate machine. The tests drive a web browser on a client machine.

With a lab environment, you can run a build-deploy-test workflow in which you can automatically build your system, deploy its components to the appropriate machines in the environment, run the tests, and collect test data. (The fully automated version of this is described in Chapter 5, "Automating System Tests.")

The workflow is controlled by a test controller with the help of test agents installed on each test machine. The test controller runs on a separate computer.

Now you might ask why you need lab environments, since you could deploy your system and tests to any machines you choose.

Well, you could, but lab environments make several things easier:

- You can set up automated build-deploy-test workflows. The scripts in the workflow use the lab role names of the machines, such as "Web Client," so that they are independent of the domain names of the computers.
- The results of tests can be shown on charts that relate them to requirements.
- Lab Manager automatically installs test agents on each machine, enabling test data to be collected. Lab Manager manages the test settings of the virtual environment, which define what data to collect.
- You can view the consoles of the machines through a single viewer, switching easily from one machine to the other.
- Lab environments manage the allocation of machines to tests for reasons that include preventing two team members from mistakenly assigning the same machine to different tests.

Lab environments come in two varieties. A *standard lab environment* (roughly equivalent to a *physical environment* in Visual Studio 2010) can be composed of any computers that you have available, such as physical computers or virtual machines running on third-party frameworks.

An SCVMM environment is made up entirely of virtual machines controlled by System Center Virtual Machine Manager (SCVMM). SCVMM environments provide you with several valuable facilities; they allow you to:

- Create fresh test environments within minutes. You can store a complete environment in a library and deploy running copies of it. For example, you could store an environment of three machines containing a web client, a web server, and a database. Whenever you want to test a system in that configuration, you deploy and run a new instance of it.
- Take snapshots of the states of the machines. For example whenever you start a test, you can revert to a snapshot that you took when everything was freshly installed. Also, when you find a bug, you can take a snapshot of the environment for later investigation.
- Pause and resume all the virtual machines in the environment at the same time.

Standard environments are useful for tests that have to run on real hardware, such as some kinds of performance tests. You can also use them if you haven't installed SCVMM or Hyper-V, as would be the case if, for example, you already use another virtualization framework. But as you can see, we think there are great benefits to using SCVMM environments.

STORED SCVMM ENVIRONMENTS

Because you can store them in a library, SCVMM environments help to make your tests repeatable; when you run them for the next build, or when a new release is planned after a six-month break, you can be sure that the tests are running under the same conditions.

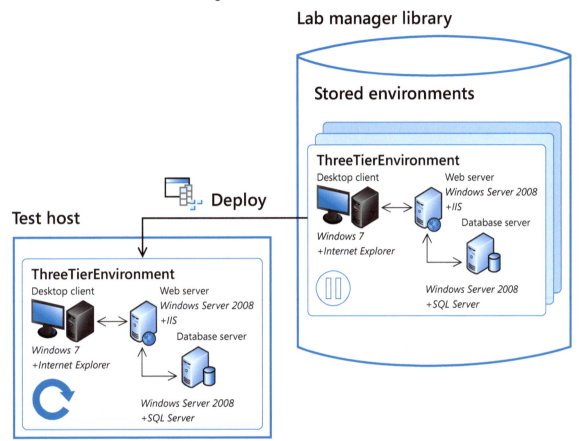

A stored SCVMM environment

For example, on Fabrikam's ice cream sales project, the team often wants to deploy and test a new build of the sales system. It has several components that have to be installed on different machines. Of course, the sales system software is a new build each time. But the platform software, such as operating system, database, and web browser don't change.

So at the start of the project, the team creates an environment that has the platform software, but no installation of the ice cream system. In addition, each machine has a test agent. The Fabrikam team stores this environment in the library as a template.

Whenever a new build is to be tested, a team member selects the stored platform environment, and chooses **Deploy**. Lab Manager takes a few minutes to copy and start the environment. Then they only have to install the latest build of the system under test.

While an environment is running, its machines execute on one or more virtualization hosts that have been set up by the system administrator. The stored version from which new copies can be deployed is stored on an SCVMM library server.

LAB MANAGEMENT WITH THIRD-PARTY VIRTUALIZATION FRAMEWORKS

Some teams have already invested in other virtualization frameworks such as VMware or Citrix Xen-Server. If that is your situation, the case for switching to Hyper-V and SCVMM might be less clear. But even if you don't install SCVMM or Hyper-V, you can still use Lab Manager by using standard environments.

With standard environments, you get many of the benefits of lab management, but without the ability to save and quickly set up fresh environments. Instead, you'd have to use your third-party machine manager to set up new machines.

When you assign a machine to a standard environment, Lab Manager will automatically install a test agent and couple it to your test controller. This makes the machine ready for an automatic build-deploy-test workflow and for test data collection. (In Visual Studio 2010, you have to install the test agent manually, but coupling it to the test controller is automatic.)

How to use lab environments

PREREQUISITES

To enable your team to use lab environments, you first have to set up:

- Visual Studio Team Foundation Server, with the Lab Manager feature enabled.
- A test controller, linked to your team project in Team Foundation Server.
- (Preferable, but not mandatory) System Center Virtual Machine Manager (SCVMM) and Hyper-V.

You only need to set up these things once for the whole team, so we have put the details in the Appendix. If someone else has kindly set up SCVMM, Lab Manager, and a test controller, just continue on here.

LAB CENTER

You manage environments by using Lab Center, which is part of Microsoft Test Manager (MTM). MTM is installed as part of Visual Studio Ultimate or Test Professional. You'll find it on the Windows Start menu under Visual Studio. If it's your first time using it, you'll be asked for the URL of your team project collection. Switch to the Lab Center view (it's the alternative to Test Center). On the Environments page, you'll see a list of environments that are in use by your team. Some of them might be marked "in use" by individual team members:

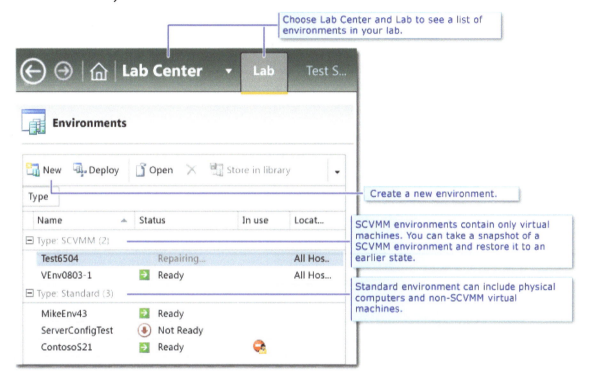

Managing environments in Lab Center

(Use the **Home** button if you want to switch to another team project.)

More information is available from the MSDN website topic: *Getting Started with Lab Management*.

CONNECTING TO A LAB ENVIRONMENT

If your team has been using lab environments for a while, then when you open Lab Center, you might already see some environments that are available to use. Pick an environment with a status of **Ready**, without an **In Use** flag, and that looks as if it has the characteristics you want, which ought to be indicated by its name. Select it and choose **Connect**.

The Environment View opens. From here you can log into any of the machines in the environment.

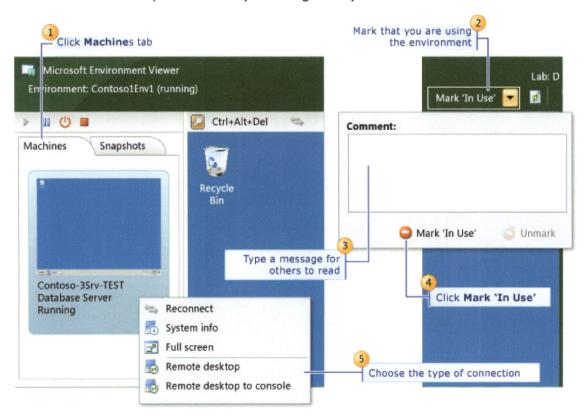

The environment view

Typically, a deployed environment will have a recent build of your system already installed. If you're sure that it's free for you to use, you could decide to run some tests on it. However, make sure you know your team's conventions; for example, if the environment's name contains the name of a team member, ask if it is ok to use.

Using a deployed (running) environment

Log in. Choose the Connect button to open a console view of the environment. From there you can log into any of its machines. More about the Connect button can be found on MSDN in the topic *How to: Connect to a Virtual Environment*.

Reserve the environment. You can mark it as **In Use** to discourage other team members from interfering with it. This doesn't prevent access by others, but simply sets a flag in Lab Center.

Revert a virtual environment to a clean snapshot. In the environment viewer, look at the Snapshots tab. If the Snapshots tab isn't available, then this is a standard environment composed of existing machines. You might need to make sure that the latest version of your system is installed.

In a virtual environment, the team member who created the environment should have made a snapshot immediately after installing the system under test. Select the snapshot and restore the environment to that state. If there isn't a snapshot available, that's (hopefully) because the previous user has already restored it to the clean state. Again, you might need to check the conventions of your team.

Explore and test your system. Now you can start testing your system, which is the topic of the next chapter.

Restore the snapshot when you're done with a virtual environment, to restore it to the newly installed state. This makes it easier for other team members to use. This option isn't available for standard environments, so you might want to clean up any test materials that you have created.

Clear the "in use" flag when you're done. Typically, a team will keep a number of running environments that contain a recent build, and share them. Reusing the environment and restoring it to its initial snapshot is the quickest way of assigning an environment for a test run.

Deploying an environment

If there is no running environment that is suitable for what you want to do, you can look for one in the library. The library contains a selection of stored virtual environments that have previously been created by your colleagues. You can learn more from the topic: *Using a Virtual Lab for Your Application Lifecycle*, on MSDN.

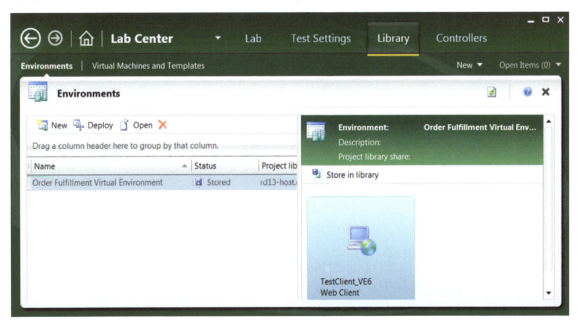

The environment library in MTM Lab Center

(If the library isn't available, that might mean that your team has not set Lab Manager to use SCVMM. But you can still create standard environments, which are made up of computers not controlled by SCVMM. Skip to the section about them near the end of this chapter. Alternatively, you could set up SCVMM as we describe in the Appendix.)

Environments stored in the library are templates; you can't connect to one because its virtual machines aren't running. Instead, you must first *deploy* it. Deploying copies the virtual machines from the library to the virtual machine host, and then starts them.

In MTM, in Lab Center, choose **Deploy**. Choose an environment from the list. They should have names that help you decide which one you want.

After you have picked an environment, Lab Center takes a few minutes to copy the virtual machines and to make sure that the test agents (which help deploy software and collect data) are running.

Eventually the environment is listed under the Lab tab as **Ready** (or **Running** in Visual Studio 2010). Then you're all set to use it. If it shows up as **Not Ready**, then try the **Repair** command. This reinstalls test agents and reconnects them to your test controller. In most cases that fixes it.

Install your system

Typically, stored environments contain installations of the base platform: operating systems, databases, and so on. They don't usually include an installation of the system under test. Your next step is therefore to install the latest build of your system.

To help choose a good recent build, open the build status report in your web browser. The URL is similar to *http://contoso-tfs:8080/tfs/web*. Click on **Builds**. You might have to set the date and other filters. The quality and location of each build is summarized.

In Lab Center, under the **Lab** tab, select the running environment and choose **Connect**. Log into the environment's machines.

Use the installer (typically an .msi file) that is generated by the build process. The location can be obtained from the build status reports. Pick an installer that was generated from the Debug build configuration. You need to put each component on the right machine. Each machine has a role name such as Client, Web Server, or Database, to help you make the right choice.

Later we'll discuss how you can write scripts to automate the deployment of the system under test.

Review the name you gave to the environment to make sure it reflects the system and build you installed.

Take a snapshot of the environment

Create a snapshot of the environment. This will enable subsequent users to get the environment back to its nice clean state. Do this immediately after you have installed your system, and before you run any tests, other than perhaps a quick smoke test to make sure the installation is OK.

You can create a snapshot either from the Snapshots tab in Environment Viewer, or from the context menu of the environment in the Lab listing.

Use it

After you've taken a snapshot, you can start using it as we described earlier. When you've finished testing, you can revert to the snapshot.

Delete it (eventually)

Delete an environment when the build it uses is superseded.

CREATING A NEW VIRTUAL ENVIRONMENT

What if there are no environments in the stored library, or none have the mix of machines you need? Then you'll have to create one. And if you're feeling generous, you could add it to the library for other team members to use.

You can either store an environment directly in the library, or you can create it as a running environment and then store it in the library. Storing directly is preferable if you don't need to configure the constituent virtual machines in any way.

To add a new environment directly to the library, open MTM; choose **Lab Center**, **Library**, **Environments**, and then the **New** command.

Creating a new environment in the library

Alternatively, to create a new running environment that you can store later, choose **Lab Center**, **Lab**, and then **New**. In the wizard, choose **SCVMM Environment**. (In Visual Studio 2010, the **New** command has a submenu, **New Virtual Environment**.)

In either method, you continue through the wizard to choose virtual machines from the library. If your team has been working for a while, there should be a good stock of virtual machines. Their names should indicate what software is installed on them.

Choose library machines that have type *Template* if they are available. Unlike a plain virtual machine, you can deploy more than one copy of a template. This is because when a template VM is deployed, it gets a new ID so that there are no naming conflicts on your network. Using templates to create a stored environment allows more than one copy of it to be deployed at a time

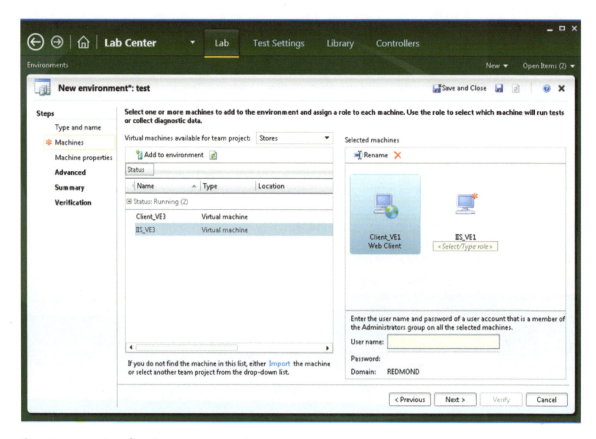

Creating a new virtual environment

You have to name each machine uniquely within your new lab environment. Notice that the name of the computer in the environment is not automatically the same as its name in the domain or workgroup.

You also have to assign a role name to each machine, such as Desktop Client or Web Server. More than one machine can have the same role name. There is a predefined set to choose from, but you can also invent your own role names. These roles are used to help deploy the correct software to each machine. If you automate the deployment process, you will use these names; if you deploy manually, they will just help you remember which machine you intended for each software component.

When you complete the wizard, there will be a few minutes' wait while VMs are copied.

MTM should now show that your environment is in the library, or that it is already deployed as a running environment, depending on what method of creation you chose to begin with. If it's in the library, you can deploy it as we described before.

After creating an environment, you typically deploy software components and then keep the environment in existence until you want to move to a new build. Different team members might use it, or you might keep it to yourself. You can mark an environment as "In Use" to discourage others from interfering with it while your work is in progress.

Stored and running machines

The lab manager library can store both individual virtual machines and complete environments. There are command buttons for creating new environments, storing them in the library, and for deploying environments from the library. You have to shut down an environment before you can store it.

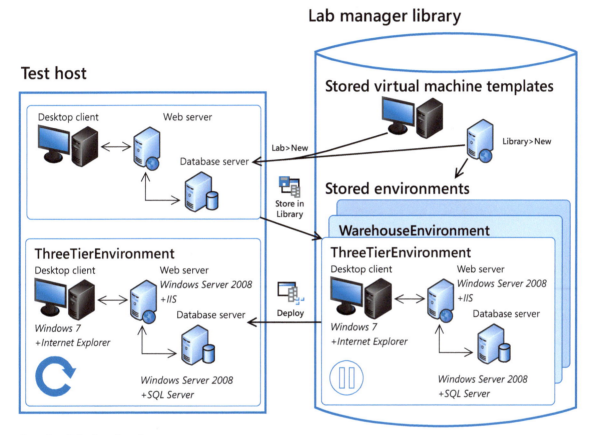

Stored and deployed environments

Creating and importing virtual machines

You can store individual virtual machines from the test host to the library. Therefore, if your team starts off with a set of virtual machines in the library that include a basic set of platforms—for example, Windows 7 and Windows Server 2008—then you can deploy a machine in an environment, add extra bits, and then store it back in the library.

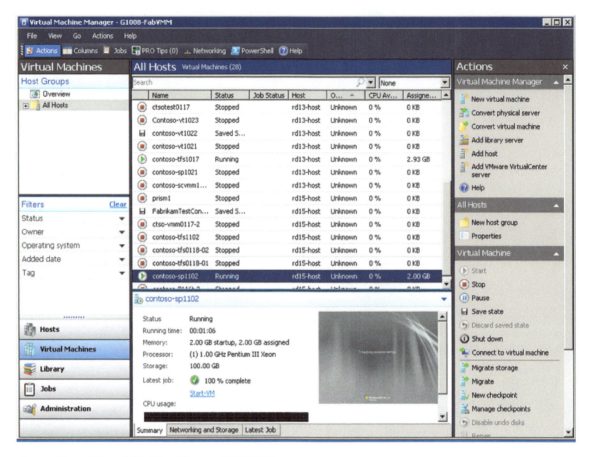

System Center Virtual Machine Manager (SCVMM)

But how do you create those first virtual machines? For this you need to access SCVMM, on which Lab Manager is based. It's typically an administrator's task, so you'll find those details in the Appendix. Briefly:

1. You can create a new machine in the SCVMM console and then install an operating system on it, either with your DVD or from your corporate PXE server.

2. Every test machine needs a copy of the Team Foundation Server Test Agent, which you can get from the Team Foundation Server installation DVD.

3. Use the SCVMM console to store the VM in the library as a template. This is preferable to storing it as a plain VM.

4. In Lab Manager, use the **Import** command on the **Library** tab in order to make the SCVMM library items visible in the Lab Center library.

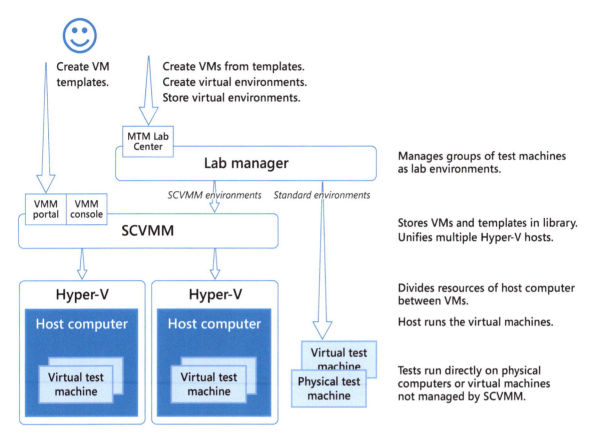

How environments are managed

COMPOSED ENVIRONMENTS

A *composed environment* is made up of virtual machines that are already running. When you compose an environment from running machines, they are assigned to your environment; when you delete the environment, they are returned to the available pool. You can create a composed environment very quickly because there is no copying of virtual machines.

We recommend composed environments for quick exploratory tests of a recent build. The team should periodically place new machines in the pool on which a recent build is installed. Team members should remember to delete composed environments when they are no longer using them.

Test host

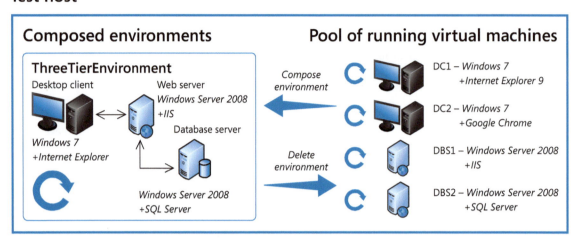

Composed environments

In Visual Studio 2012, you make a composed environment the same way you create a virtual environment: by choosing **New** and then **SCVMM environment**. In the wizard, you'll see that the list of available machines includes both VM templates and running pool machines. If you want, you can mix pool machines and freshly created VMs both in the same environment. For example, you might use new VMs for your system under test, and a pool machine for a database of test data, or a fake of an external system. Because the external system doesn't change, there is no need to keep creating new versions of it.

In Visual Studio 2010, use the **New Composed Environment** command and choose machines from the list.

STANDARD ENVIRONMENTS

Standard environments are made up of existing computers. They can be either physical or virtual machines, or a mixture. They must be domain-joined.

You can create standard environments even if your team hasn't yet set up SCVMM. For example, if you are already using VMware to run virtual machines and don't want to switch to Hyper-V and SCVMM, you can use Lab Manager to set up standard environments. You can't stop, start, or take snapshots of standard environments, but Lab Manager will install test agents on them and you can use them to run a build-deploy-test workflow.

You can also use standard environments when it is important to use a real machine—for example, in performance tests.

To create a standard environment, click **New** and then choose **Standard Environment**.

(In Visual Studio 2010, choose **New Physical Environment**. You must manually install test and lab agents on the computers. These agents can be installed from the Team Foundation Server DVD.)

For an example, see *Lab Management walkthrough using Visual Studio 11 Developer Preview Virtual Machine* on the Visual Studio Lab Management team blog.

Summary

There's a lot of pain and overhead in configuring physical boxes to build test environments. The task is made much easier by Visual Studio Lab Manager, particularly if you use virtual environments.

With Lab Manager you can:

- Manage the allocation of lab machines, grouping them into lab environments.
- Configure machines for the collection of test data.
- Rapidly create fresh virtual environments already set up with a base platform of operating system, database, and so on.

Differences between Visual Studio 2010 and Visual Studio 2012

- **System Center Virtual Machine Manager 2012**. Lab Management in Visual Studio 2012 works with *SCVMM 2012* in addition to SCVMM 2008.
- **Standard environments**. Lab Manager in Visual Studio 2012 is easier to use with third-party virtualization frameworks as well as physical computers. It will install test agents if necessary.
- **Test agents**. In Visual Studio 2010, you must install test and lab agents on the machines that you want to use in the lab. In Visual Studio 2012, there is only one type of agent, and it is installed automatically by Lab Manager on each of the machines in a lab environment. You can still install the test agent yourself to save time when lab environments are created.
- **Compatibility**. Most combinations of 2010 and 2012 products work together. For example, you can create environments on Visual Studio Team Foundation Server 2010 using Microsoft Test Manager 2012.

Where to go for more information

All links in this book are accessible from the book's online bibliography available on MSDN: *http://msdn.microsoft.com/en-us/library/jj159339.aspx.*

4   Manual System Tests

Manual testing is as old as computer programming. After all, most systems are designed to be used by someone.

Anyone can try out a system, just by using it. But testing it fully is an engineering discipline. An experienced tester can explore an application systematically and find bugs efficiently; and conversely, can provide good confidence that the system is ready to be released.

In this chapter, we'll discuss tools and techniques for verifying that your software does what its users and other stakeholders expect of it. We'll show how to relate your tests to requirements (whether you call them user stories, use cases, or product backlog items). You will be able to chart the project's progress in terms of how many tests have passed for each requirement. And when requirements change, you can quickly find and update the relevant tests.

When a bug is found, you need to be able to reproduce it. This can be one of the biggest sources of pain in bug management. We'll look at tools and techniques that help you trace how the fault occurred. And when it's fixed, you can retrace your steps accurately to make sure it's really gone.

When members of the traditional development teams at Contoso speak of testing, manual system testing is usually what they mean. Unlike unit tests, they can't be run overnight at the touch of a button. Running a full set of system tests takes as long as it takes someone to exercise all the system's features. This is why traditional companies like Contoso don't release product updates lightly. No matter how small an update, they have to run all the tests again, just in case the update had unintended consequences for another feature. And as the project goes on and the system gets bigger, a full test run gets longer.

In later chapters, this will be our motivation for automating system tests. However, the testing tools in Visual Studio include a number of ways to speed up manual retesting. For example, you can use test impact analysis to focus just on the tests that have been affected by recent changes in the code. Another way is to record your actions the first time around, and replay them the next time: all you have to do is watch the playback happening and verify the results.

We will never drop manual testing completely. Automated tests are excellent for regression testing—that is, verifying that no faults have developed since the last test—but are not so good for finding new bugs. Furthermore, there's a tradeoff between the effort required to automate a test and the costs of rerunning it manually. Therefore, we'll always do manual testing when new features are developed, and most projects will continue to perform at least some of their regression tests manually.

Microsoft Test Manager supports manual system tests

Microsoft Test Manager (MTM) is the client application that supports the testing features of Visual Studio Team Foundation Server. Get it by installing Visual Studio Ultimate or Visual Studio Test Professional.

In MTM, you can run tests in two modes: exploratory and scripted test cases. In exploratory testing you run the system and see what you can find. With test cases, you work through a script of steps that you either planned in advance, or that you worked out while you were exploring.

Exploratory testing is a lightweight and open approach to testing: nothing is prescribed, except that you might want to focus on a particular user story. Scripted test cases are more repeatable, even by people who aren't familiar with the application.

Exploratory testing

Art, one of the more senior people at Contoso, might very well ask, "I've been doing exploratory testing for thirty years. I never needed any special tools. You just run the system and try it out. Why do I need Microsoft Test Manager?"

Well, that's true, you don't need it. But we think it can make several aspects of system testing less troublesome and faster, such as reporting bugs. Let's try it and see.

We'll start with a scenario that's very easy to set up and experiment with.

Let's assume that you want to test a website that already exists. You are at the start of a project to improve its existing features as well as to add some new ones. Your objective in this testing session is to find any actual bugs, such as broken links, and also to look for any places where things could be made better. Also, you just want to familiarize yourself with the existing website.

On your desktop machine you have installed Microsoft Test Manager, and you have a web browser such as Internet Explorer.

Art would just start up the web browser and point it at the website. Instead, you begin by putting Microsoft Test Manager into exploratory mode. This allows it to record what you do and makes it easy for you to log bugs.

1. Open Microsoft Test Manager. You might have to choose a team project to log into, and you might have to select or create a test plan.

2. Choose **Testing Center**, **Test**, **Do Exploratory Testing**, and finally **Explore**. (If you have specific requirements that you are testing, you can select the requirement and choose **Explore Work Item**. Otherwise, just choose **Explore**.)

MTM Testing Center

The Testing Center window minimizes and Microsoft Test Runner (the exploratory testing window) opens at the side of the screen. In the new window, click **Start to start recording your actions**.

Exploratory testing in Microsoft Test Runner

Now open your web browser and point it at the website you want to test.

As you work, you can write notes in the testing window, insert attachments, and take screenshots. After you take a screenshot, you can double click to edit it so as to highlight items of interest.

Making notes in Test Runner

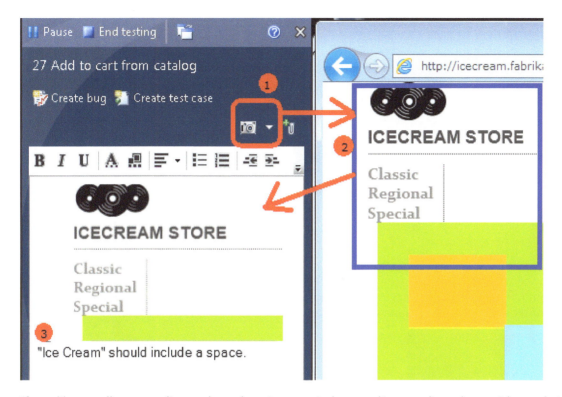

If you like to talk your audience through an issue, switch on audio recording, along with a real-time recording of what happens on the screen.

In addition, Test Runner records your keystrokes, button clicks, and other actions as you work. If you create a bug report or a test case, your recent actions are automatically included in the report in readable form, and can also be played back later.

If you want to do something else partway through a test, suspend recording by using the **Pause** button.

> **Tip:** *If you receive a distracting email or instant message during your test session, click Pause before you type your reply. The same applies if you get a sudden urge to visit an off-topic website. You don't want your extramural interests to be circulated among your team mates as part of a bug report.*

CREATING BUGS

If you find an error, just click **Create Bug**. The bug work item that opens has your recent actions already listed, along with your notes and other attachments. Choose **Change Steps** if you don't want to include all the actions you took from the start of your session.

Edit the bug, for example, to make sure that it has the correct iteration and area for your project, or to add more introductory text, or to assign it to someone. When you save the bug, it will go into the team project database and appear in queries and charts of currently open bugs.

Notice that the bug has a Save and Create Test button. This creates a test case that is specifically intended to verify that this bug has been fixed. The test case will contain a script of the same steps as the bug report, and the two work items will be linked. After the bug has been fixed, this test case can be performed at intervals to make sure that the bug does not recur.

CREATING TEST CASES

At any point while you are exploring, you can create a test case work item that shows other testers how to follow your pioneering steps. Click **Create Test Case**. The actions that you performed will appear as a script of steps in the test case. Choose **Change Steps** to adjust the point in the recording at which the test case starts, and to omit or change any steps. Then save the test case.

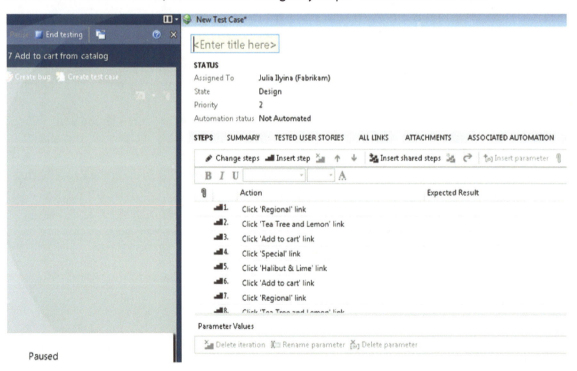

New test case

If you started your exploration in the context of a particular requirement work item, the test case will be attached to it. On future occasions when this requirement is to be tested, testers can follow the script in the test case.

> **Tip:** *Consider creating test cases or bug reports in each exploratory session. If you find a bug, you want to report it. If you don't find a bug, you might want to create test cases. These make it easy to verify that future builds of the system are still working. The test case contains your steps so that anyone can easily follow them.*

For more about how to do exploratory testing, see the topic on MSDN: *Performing Exploratory Testing Using Microsoft Test Manager.*

No more "no repro"

So what do we get by performing exploratory testing with Microsoft Test Manager rather than just running the system?

One of the biggest costs in bug management is working out exactly how to make the bug appear. Bug reports often have inaccurate or missing information. Sometimes, in a long exploration, it's difficult to recall exactly how you got there. Consequently, whoever tries to fix the bug—whether it's you or someone else—can have a hard time trying to reproduce it. There's always the suspicion that you just pressed the wrong button, and a lot of time can be wasted passing bugs back and forth. With action recording, there's much less scope for misunderstanding.

It gets even better. When we look at using lab environments, we'll see how execution traces and other diagnostic data can be included in the bug report, and how developers can log into a snapshot of the virtual environment at the point at which the bug was found.

But first let's look at test cases.

Testing with test cases

Test cases are the alternative to exploratory testing. In exploratory testing, you go where your instincts take you; but a test case represents a particular procedure such as "Buy one tub of pink ice cream." You can, if you want, plan a detailed script that you follow when you run the test case.

A test case is a specific instance of a requirement (or user story or product backlog item or whatever your project calls them). For example if the requirement is "As a customer, I can add any number of ice creams to my shopping cart before checking out" then one test case might be "Add an oatmeal ice cream to the cart, then check out" and another might be "Add five different ice creams."

Test cases and requirements are both represented by work items in your team project, and they can (and should) be linked.

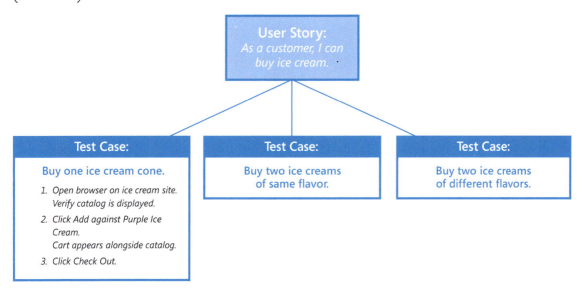

One requirement typically has several test cases

Test cases are typically created in two situations:

- Before coding. When your team reviews a requirement in preparation for implementing it. Typically this is at the start of a sprint or iteration. Writing test cases is a great way of nailing down exactly what a requirement means. It forms a clear target for the developers. When the code has been written, the requirement isn't complete until its test cases pass.
- After coding. When you do exploratory testing after the requirement has been implemented, you can generate a test case to record what you did. Other testers can quickly repeat the same test on future builds.

CREATING TEST CASES BEFORE CODING THE REQUIREMENTS

Test cases are a good vehicle for discussing the exact meaning of a requirement with the project stakeholders. For example, another test case for that "I can add any number of ice creams" requirement might be "Don't add any ice creams; just go to check out." This makes it obvious that there might be something wrong with this requirement. Maybe it should have said "one or more."

> **Tip:** *Inventing test cases is a good way of taking the ambiguities out of requirements. Therefore you should create test cases for all requirements before they are implemented. Discussing test cases should be a whole-team activity, including all the stakeholders who have a say in the requirements. It isn't something the test lead does privately.*

To add test cases in Microsoft Test Manager, choose **Testing Center**, **Plan**, **Contents**.

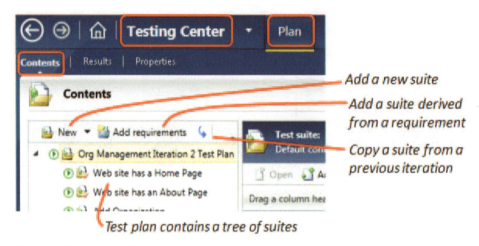

The test plan can contain test suites

Select the root test plan, and choose **Add Requirements**. This button opens a Team Foundation Server query that will find all the requirements in the project. Before running the query, you might want to add a clause to narrow it down to the current iteration.

Select the items you want (CTRL+A selects all) and choose **Add requirements to Plan**.

A test suite is created for each requirement, with a single test case in each suite. A suite is a collection of test cases that are usually run in the same session.

You'll probably want to add more test cases for each requirement. A suite of this kind remembers that it was created from a requirement; when you create new test cases in it, they are automatically linked to the requirement.

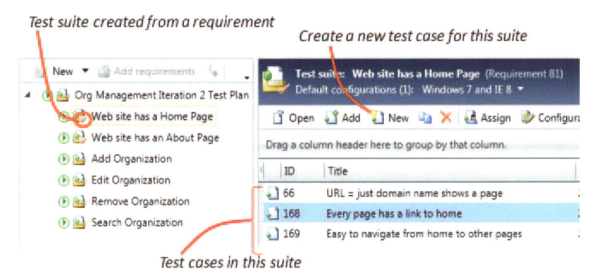

Test suite created from a requirement

Create a new test case for this suite

Test cases in this suite

Test cases automatically linked to requirement

Discussions about the requirements often result in new ones. You can create new requirements by using the **New** menu near the top right of the window.

Test cases have steps

Test cases usually contain a series of steps for the tester to follow. They can be very specific—enter this text, click that button—or much more general—Order some ice cream. With specific instructions, a tester who does not know the application can reliably perform the test. With more general instructions, there is more room for the tester to explore and use her own ingenuity to break the system.

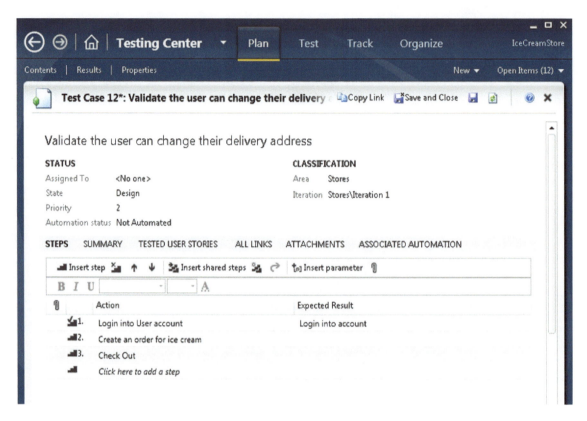

Test case steps

For each sequence of steps that you want to write, create a new test case in the test suite that is derived from the requirement.

Later in this chapter, we'll discuss the process of inventing test cases and test case steps.

CREATING A TEST CASE AFTER THE CODE IS WRITTEN

We've already seen how you can perform an exploratory test and create a test case from the action recording. You can add the test case to the relevant requirements after creating it. Alternatively, when you start an exploratory test, you can select the requirement that you intend to investigate; by default, any bug or test case you create will be linked to that requirement.

RUNNING A TEST CASE

In Microsoft Test Manager, choose **Testing Center**, **Test**, **Run Tests**, and then select a suite or one or more individual tests and choose **Run**.

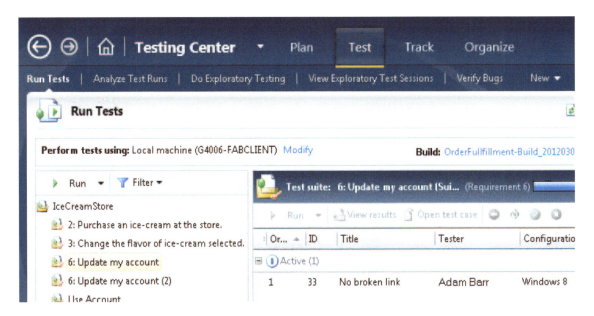

Run tests

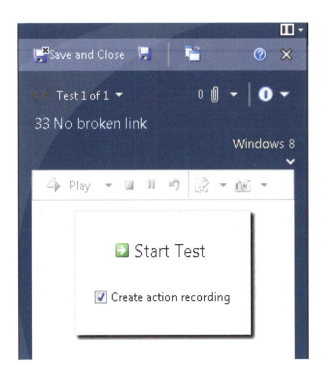

Microsoft Test Runner start screen

Microsoft Test Runner opens at the side of the screen, as seen above.

Check **Create action recording** to record your steps so that they can be played back rapidly the next time you run this test.

Make sure your web browser is ready to run. Click **Start Test**.

Work through the steps of the test case. Mark each one Pass or Fail as you go.

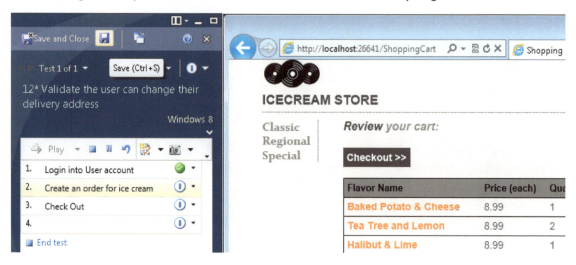

Starting a test

When you complete the test case, mark the whole test as Pass or Fail. You can also mark it Blocked if you were unable to conclude the test. Save and close the test case to store the result.

The result you record will contribute to the charts that are visible in the project web portal. You can also run queries to find all the test cases that have failed.

Recording your results and logging a bug
In addition to your pass/fail verdict, you can attach to the bug report your comments, files, snapshots of the screen, and (if you are using a virtual lab environment) snapshots of the state of the environment. You might have to pull down the menu bar continuation tab to see some of these items.

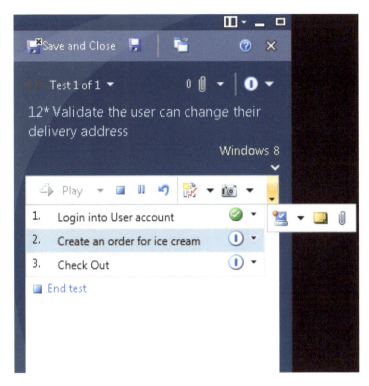

Validating a step

If you find a fault, create a bug work item:

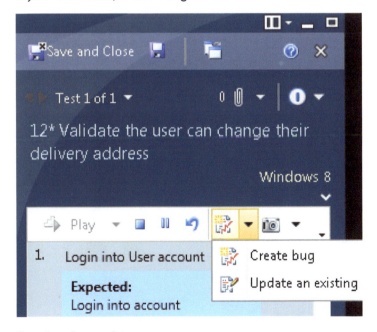

Creating a bug work item

The new bug will automatically contain a record of the steps you took, as well as your comments and attachments. You can edit them before submitting the bug.

REPLAYING ACTIONS ON LATER RUNS

The first time you run a test case, you can record your actions—button clicks, typing into fields, and so on. When you rerun the test case, you can replay the actions automatically, either one step at a time, or the whole run.

This is very useful for two purposes:

- Bug replay. Anyone who investigates a bug you reported can rerun your test and see exactly what you did and where it went wrong. Any bugs you logged are linked to the test case, so it's easy to navigate from the bug to the test case and run it.
- Regression tests. You (or any other team member) can rerun the test on any future build, to verify that the test still passes.

To replay a test for either purpose, open the test case in Microsoft Test Manager, and choose the most recent run. Choose **Run**, and then in the test runner choose **Start Test**. (Do not check **Overwrite existing action recording**.) Choose **Play**. The actions you recorded will be replayed.

This is a great facility, because it can get the developer to the point where you found the bug. But be aware that the recording isn't perfect. Some actions, such as drawing on a canvas, aren't recorded. However, button clicks and keystrokes are recorded correctly.

> **Tip:** *Action recording relies on each input element having a unique ID. Make sure the developers know this when designing both HTML and Windows desktop applications.*

BENEFITS OF TEST CASES IN MICROSOFT TEST MANAGER

What have we gained by using test cases?

- **No more "no repro."** Just as with exploratory testing, bugs automatically include all the steps you took to get to the bug, and can include your notes and screenshots as well. You don't have to transcribe your actions and notes to a separate bug log, and you don't have to recall your actions accurately.

- **Test cases make requirements runnable.** Requirements are typically just statements on a sticky note or in a document. But when you create test cases out of them, especially if there are specific steps, the requirement is much less ambiguous. And when the coding is done and checked in, you can run the test cases and decide whether the requirement has been met or not.

- **Traceability from requirements to tests.** Requirements and test cases are linked in in Team Foundation Server. When the requirements change, you can easily see which tests to update.

- **Rapid and reliable regression testing.** As the code of the system develops, it can happen that a feature that worked when it was first developed is interfered with by a later update. To guard against this, you want to rerun all your tests at intervals. By using the action replay feature, these reruns are much less time consuming than they would otherwise be, and much less boring for the testers. Furthermore, the tests still produce reliable results, even if the testers are not familiar with the system.

- **Requirements test status chart.** When you open the Reports site from Team Explorer, or the Project Portal site, you can see a report that shows which requirements have passed all their associated tests. As a measure of the project's progress, this is arguably more meaningful than the burndown chart of remaining development tasks. Powering through the work means nothing unless there's also a steady increase in passing system tests. Burndown might help you feel good, but you can't use it to guarantee good results at the end of the project.

Title	Work Progress			Test Status	
	% Hours Completed	Hours Remaining	Test Points	Test Results	Bugs
Customers can buy ice cream.	60 %	16	3	67 % / 33 %	3
Customers can select flavor from catalog.	60 %	10	3	67 % / 33 %	
Vendor can vary flavor catalog.	25 %	15	2	100 %	
Vendor can set different prices for flavors.	39 %	23	2	100 %	4
Customers can set favorite flavors.	100 %	0	1	100 %	4
Users can choose UK spelling of favourite flavour.	100 %	0	6	50 % / 33 %	
Customers can select different types of cones.	75 %	5	22	63 % / 26 %	2

Requirements test status chart

How test cases are organized

TEST PLANS

When you first start MTM, you are asked to choose your team project, and then to choose or create a test plan. You always run tests in the context of a test plan. Typically you have a separate test plan for each area and iteration of your project. If, later, you want to switch to a different project or plan, click the **Home** button.

A test plan binds together a set of test cases, a particular build of your product, test data collection rules, and the specification of lab environments on which the plan can be executed. Test cases are referenced by a tree of test suites. A suite is a group of tests that are usually run in the same session. A test case can be referenced from more than one test suite and more than one test plan.

Faults in your system can be reported using Bug work items, which can be linked to particular requirements. A test case can be set up to verify that a particular bug has been fixed.

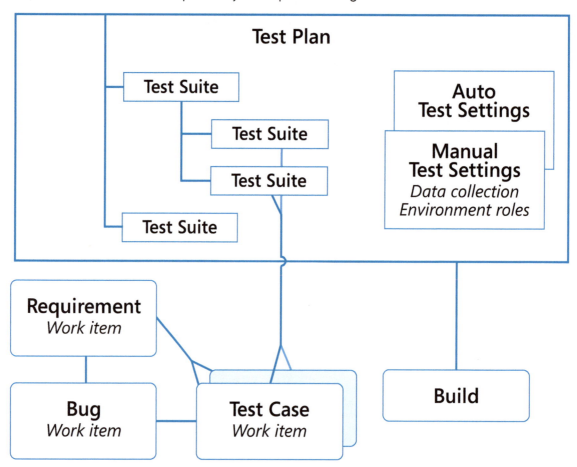

Test plans contain suites, which refer to test cases

Test suites

A test plan contains a tree of *test suites*, and each suite contains a list of test cases. A test case can be in any number of test suites. A suite is a group of tests that are usually run in the same session.

You can make a tree structure of nested suites by using the **New**, **Suite** command. You can drag suites from one part of the tree to another.

There are two special types of suites:

- Requirements-linked suites, which we have already encountered.
- Query-based suites. You sometimes want to run all the test cases that fulfill a particular criterion—for example, all the Priority 1 tests. To make this easy, choose **New**, **Query-based suite**, and then define your query. When you run this suite, all the test cases retrieved by the query will be run.

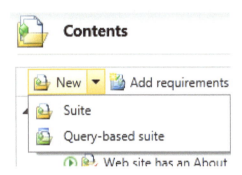

Choosing new query-based suite

Shared steps

Certain sub-procedures are common to many tests. For example, opening an application on a particular file or logging in. Shared step sets are like subroutines, although they can't be nested.

To create a set of shared steps, select a contiguous subset of steps in a test case, and choose **Create Shared Steps**. You have to give the set a name. In the test case, the set is replaced by a link.

To reference a set of shared steps that already exists, choose **Insert shared steps**.

When the test case is run, the shared steps will appear in line in the test runner.

PARAMETERS

You can set parameters to make generic test cases. For example, you could write the instruction in a step as "Open file @file1." The test case thereby acquires a parameter @file1, for which you can provide values.

When you define parameters, a matrix appears in which you can set combinations of parameter val-

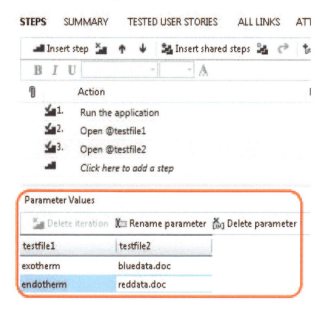

ues:

Set parameter values during test planning

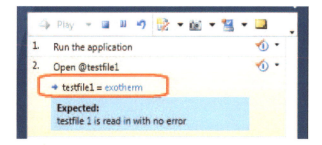

When you run the test case, the steps appear with parameter values:

Expected values displayed when you run the test

When you complete one run, you are presented with another, in which the parameters are set to the next row in the table of values.

Configurations

Most applications are expected to run on a variety of different versions and implementations of operating systems, browsers, databases, and other platforms.

When you design a test case—either in code or by defining manual steps—you design it to work with a particular configuration. For example, you might write a test case under the assumption that it will be running on Windows Server 2008, where the steps would be different than under, say, Linux. Or your test might include starting up a particular utility; in other words, you are assuming that it will be available when the test is run.

When you design a test case, you can record these assumptions as Configuration properties of the test case.

When you want to run a test case, you can filter the list of tests by configuration, to match the platform and applications that you have available.

To set the configuration properties of a test case, choose **Plan**, select the test case, and choose **Configurations**:

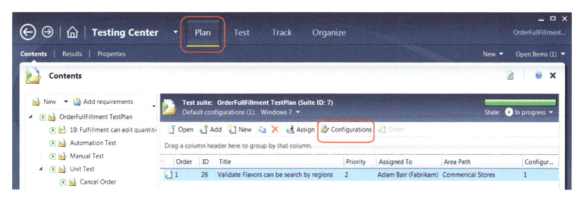

Setting configurations

To define your own configurations in addition to the built-in sets, choose **Testing Center**, **Organize**, **Test Configuration Manager**.

To filter the list of tests by configuration to match the applications that you have on your test machine, choose **Testing Center**, **Test**, and then set the required **Filter**. You can either set the filter for the test plan as a whole, or for the test suite.

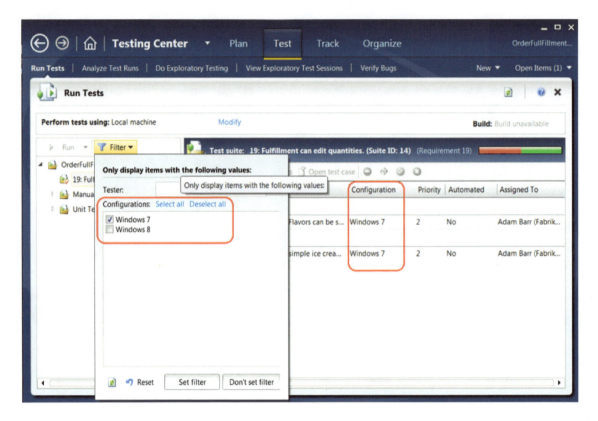

Setting test configurations

For more information about test configurations in Visual Studio 2010, see *Defining Your Test Matrix Using Test Configurations*.

Build

You can also define which build you are using for the test plan. This is just a matter of documenting your intentions: it doesn't automatically deploy that build, but the build id will appear in reports. It also appears prominently in the plan, to remind testers to obtain that particular installer.

To set a build, choose **Plan**, **Properties**, and under Builds, **Modify**.

The drop-down menu shows all the server builds that have been successfully performed recently, as described in the previous chapter.

Testing in a lab environment

In the previous sections of this chapter, we made the assumption that the system we're testing is a live website, which meant we could focus on features like action recording and playback, and avoid issues like installation and data logging. Let's now move to a more realistic scenario in which you are developing a website that is not yet live. To test it, you first have to install it on some computers. A lab environment, which we learned all about in the previous chapter, is ideal.

Why not just install your system on any spare computer?

• As discussed earlier, lab environments, particularly virtual environments, can be set up very quickly. Also the machines are newly created, so that you can be certain they have not been tainted by previous installations.

• If we link the test plan to the lab environment, Microsoft Test Manager can collect test data from the machines, such as execution traces and event counts. These data make it much easier to diagnose the fault.

• When you find a bug, you can snapshot the state of the virtual environment. Whoever will try to fix the bug can log into the snapshot and see its state at the point the bug was found.

CLIENT OUTSIDE THE LAB: TESTING WEB SERVERS

Let's assume you are testing a web server such as a sales system. We'll put the web server in the lab environment. But just as when we were testing the live website, your desktop computer can be the client machine, because it only has to run a web browser—there is no danger to its health from a faulty installation.

Create the lab environment

Switch to the Lab Manager section of MTM, and set up a virtual environment containing the server machines that your system requires, as we described in the previous chapter.

> **Tip:** *If one of the machines is a web server, don't forget to install the **Web Deployment Tool** on it. Also, open Internet Information Services Manager. Under Application Pools, make sure that your default pool is using the latest .NET Framework version.*

The status of the environment must be **Ready** in order for MTM to be able to work with it. (If it isn't, try the **Repair** command on the shortcut menu of the environment.)

If you perform tests of this system frequently, it's useful to store a template for this environment.

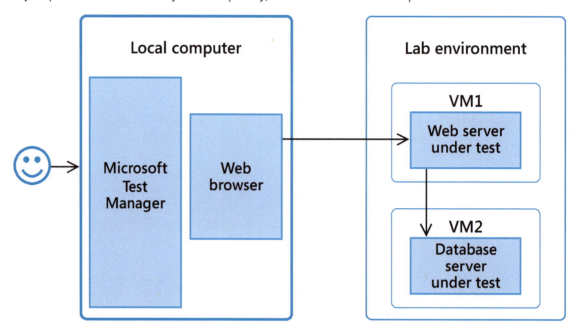

Environment template

Tell the test plan about the environment

After creating a lab environment, configure the test settings of your test plan.

In MTM, select **Testing Center**, **Plan**, **Properties**. Under **Manual Runs**, select your **Test Environment**. Make sure that **Test Settings** is set to <Default>. (You can create a new test setting if you want a specific data collection configuration.)

This enables Microsoft Test Manager to collect test data from the machines in the lab. (More about these facilities in Chapter 6, "A Testing Toolbox.")

Get and install the latest build

To find the latest build of your system, open the Builds status report in your web browser. The URL is similar to http://contoso-tfs:8080/tfs. Choose your project, and then **Builds**. Alternatively, open Team Explorer on your project and choose **Builds**. The quality of each recent build is listed, together with the location of its output. Use the installers that should be available there: you should find .msi files.

If no build is available, you might need to configure one: see Chapter 2, "Unit Testing: Testing the Inside." More fundamentally, you might need to set up the build service: see Appendix, "Setting up the Infrastructure."

Connect to each machine in the virtual environment and install the relevant component of your system using the installers in the same way that your users will. Don't forget to log a bug if the installers are difficult to use.

> **Tip:** *After you have installed a build, take a snapshot of the environment so that it can be restored to a clean state for each test.*

In the next chapter we'll look at automating deployment, which you can do whether or not you automate the tests themselves.

Installing a web service

Web services are a frequent special case of installation. Provided you have installed the Web Deploy tool (MSDeploy) on the machine running Internet Information Services (IIS), you can run the installation from any computer. Run the deploy command from the installation package, providing it with the domain name of the target machine (not its name in the environment) and the credentials of an administrator:

```
IceCream.Service.deploy.cmd /y /m:vm12345.lab.contoso.com /u:ctsodev1 /p:Pwd
```

> **Tip:** *Run the command first with the /T option and not /Y. This verifies whether the command will work, but does not actually deploy the website.*

Start testing

In Microsoft Test Manager, switch back to Testing Center and run the tests. Don't forget that when you start your web browser, you will have to direct it to the virtual machine where the web server is running.

On finding a bug, save the virtual environment

If the developer's machine isn't a perfect clone of your environment, what didn't work in your test might work on hers. So, can we reproduce your configuration exactly?

Yes we can. Use the environment snapshot button to save your virtual environment when you log the bug. The developer will be able to connect to the snapshot later.

This allows you to reproduce the bug substantially faster and more easily.

Be aware that anyone who opens the snapshot will be logged in as you, so don't make it readable by anyone who might think it's a great joke to write indelicate messages to your boss on your behalf. Many teams use different accounts for running tests.

Client in the lab: testing desktop and thick-client apps

The previous lab configuration relied on the idea that the only interesting state was on the server machines, so only they were in the lab environment. But if you are testing a desktop application, or a web application with a significant amount of logic on the client side, then:

- If you save the state of the environment for bug diagnosis, you want this to include the client computer.
- The application or its installer could corrupt the machine. It is therefore preferable to install and run it on a lab machine.

For these reasons, an application of this type should run entirely on the lab environment, including the thick client or stand-alone application.

However, you must also install MTM on the client computer. To be able to record and play back your actions, MTM has to be running on the same computer as the user interface of the system under test.

This leads to a slightly weird configuration in which one copy of MTM is installed inside the lab environment to run your tests, and another copy is on your desktop machine to manage the environment and plan the tests.

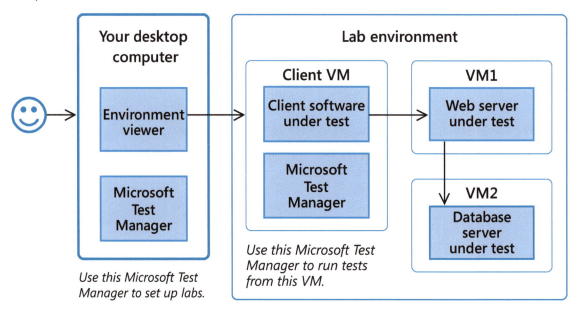

Using MTM to test client software

If you often create environments with this setup, it is a good idea to store in your VM Library a machine on which MTM is installed.

Test impact analysis

Test impact analysis recommends which tests should be run once more, based on which parts of the code have been updated or added.

As the project progresses, you'll typically want to focus your testing efforts on the most recently implemented requirements. However, as we've noted, the development of any feature often involves updates in code that has already been written. The safest—perhaps we should say the most pessimistic—way to make sure nothing has been broken is therefore to run all the tests, not only for new features, but for everything that has been developed so far.

However, we can do better than that. An important feature of the testing toolbox in Visual Studio is test impact analysis (TIA).

To use TIA, you have to set various options, both in the build process and in the properties of the test plan, as we'll explain in a moment. As you run tests, the TIA subsystem makes a note of which lines of code are exercised by each test case. On a later occasion, when you use a new build, TIA can work out which lines of code have changed, and can therefore recommend which tests you should run again.

Notice that TIA will recommend only tests that you have run at least once before with TIA switched on. So when you create new test cases, you have to run them at least once without prompting from TIA; and if you switch on TIA partway through the project, TIA will only know about tests that you run subsequently.

TIA ignores a test case if it fails, or if a bug is generated while you run it. This ensures that, if it recommends a test case that fails, it will recommend it again. However, it also means that TIA will not recommend a test until it has passed at least once.

Enabling test impact analysis

To make TIA work, you have enable it in both the build definition and in your test plan, and then create a baseline.

1. **Enable TIA in the build that you use for manual testing**; that is, the periodic build service that generates the installers that you use to set up your system. We discussed setting up a build in Chapter 2, "Unit Testing: Testing the Inside."
   In Visual Studio Premium or Ultimate, connect to the team project, then open the build definition. On the **Process** tab, set **Analyze Test Impact** to **True**.
   Queue a build for that definition.

2. **Enable TIA in your test plan**. In MTM, connect to your team project and test plan, and then open the **Plan**, **Properties** tab. Under **Manual runs**, next to **Test settings**, choose **Open**. In the test plan settings, on the **Data and Diagnostics** page, enable the **Test Impact** data collector.
   If you use a lab environment, you enable it on each machine. For IIS applications, you also need to enable the ASP .NET Client Proxy data collector.

3. **Create a baseline**. TIA works by comparing new and previous builds.
   a. **Deploy your application** by using installers from a build for which you enabled TIA.
   b. **Specify which build you have deployed and run your tests**:
      In Microsoft Test Manager, choose **Testing Center**, **Test**, **Run Tests**. Choose **Run with Options** and make sure that **Build in Use** is set to the build that you have installed.

Using test impact analysis

Typically, you would run test impact analysis when a new major build has been created.

In your test plan's properties, update **Build in use** to reflect the build that you want to test.

When you want to run test cases, in MTM, choose **Testing Center**, **Track**, **Recommended Tests**. Verify that **Build to use** is the build you will deploy to run your tests. Select **Previous build to compare**, then choose **Recommended tests**.

In the list of results, select all the tests and choose Reset to active from the shortcut menu. This will put them back into the queue of tests waiting to be run for this test plan.

Devising exploratory tests and scripted test cases

Read James A. Whittaker's book *Exploratory Software Testing* (Addison-Wesley Professional, 2009). In it, he describes a variety of tours. A tour is a piece of generalized advice about procedures and techniques for probing a system to find its bugs. Different tours find different kinds of bugs, and are applicable to different kinds of systems. Like design patterns, tours provide you with a vocabulary and a way of thinking about how to perform exploratory testing. As he says, "Exploration does not have to be random or ad hoc."

Testing tours and tactics are sometimes considered under the headings of positive and negative testing, although in practice you'll often mix the two approaches.

Positive testing means verifying that the product does the things that it is supposed to under ordinary circumstances. You're trying to show that it does what's on the tin. On the whole, detailed test case scripts tend to be for positive tactics.

Negative testing means trying to break the software by giving it inputs or putting it into states that the developers didn't anticipate. Negative testing tends to find more bugs. Negative testing is more usually done in exploratory mode; there are also automated approaches to negative testing.

MOSTLY POSITIVE TACTICS

Storyboards

A storyboard is a cartoon strip—usually in the form of a PowerPoint slide show—that demonstrates an example of one or more requirements being fulfilled from the stakeholders' point of view. It doesn't show what's going on inside the software system; but it might show more than one user.

Each slide typically shows a mock-up of a screen, or it shows users doing something.

The purpose is to refine the details of the system's behavior, in a form that is easy to discuss, particularly with clients who are expert in the subject matter but not in software development.

Here is a storyboard for part of the ice-cream website:

A storyboard

There would be other storyboards for the production manager to add new flavors and change prices, for dispatchers to log in, and so on.

Stories can be told at different levels. Some storyboards are at the epic level, telling an end-to-end story in which users' overall goals are satisfied. The customer gets the ice cream, the vendor gets the money. Others are at the level of the typical user story, describing features that make up an overall epic. For example: As a customer, I can add any flavor in the catalog to my shopping cart. The storyboard might show a mock-up of the catalog display.

It's traditional in storyboards to use a font that makes it look as though you drew these slides casually on the whiteboard. Which of course, you can also do. But if you have Visual Studio 2012, you get a PowerPoint add-in that provides you with mock-ups of user interface elements, so that you can make the whole thing look as though it has already been built. (Take care when showing it to prospective users; they will think it already exists.)

Storyboards and test cases

The storyboard can be the basis of a test case in which the slides represent the steps. Link the storyboard file to the test case work item. If someone changes the storyboard, they can see that the test case should be updated.

Like storyboards, test cases can be written at different levels. Remember to create test cases at the epic end-to-end level, as well as at the detailed level.

Storyboards typically illustrate the "happy path," leaving aside the exception cases. Don't forget to test for the exceptions! To make sure the exceptions get tested, add test cases for them. It isn't always necessary to write down the steps of each test case, and unless an exceptional case is complex, just the title will do. Alternatively, you might just leave exceptions to exploratory testing.

Create, read, update, and delete (CRUD)

It's important to make sure that you've covered all the application's features in your tests. Here's a systematic technique for doing that, in which you sketch an entity-relational diagram, and then check that you have tests for updating all the elements on the diagram.

Look at the entities or objects that you see in the user interface as you use the system. Ignoring all the decorative material, what important objects are represented there, and what relationships are there between them? For example, maybe you can browse through a catalog of ice cream flavors. OK, so let's draw boxes for the Catalog, and some example Flavors and relationships between them. By choosing items from the catalog, you can fill a shopping cart with different flavors. So, now we have a Cart, but let's call it an Order, since that's what it becomes if we choose to complete the purchase. And when you want to complete your purchase, you enter your name and address, so we have a Customer connected to the Order:

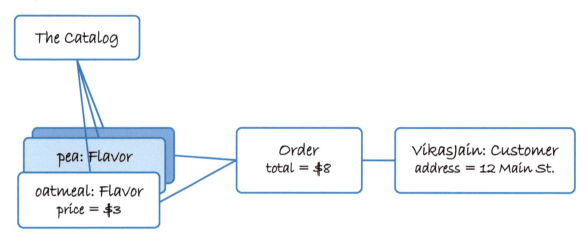

Instance diagram of an example order

The diagram doesn't have to correspond to anything in the system code. These entities just represent the concepts you see in the system's behavior. They are written in the language of the users, not the code. The diagram helps clear away the inessential aspects of the user interface, so that you can concentrate on the important things and relationships.

Look at each attribute and consider what a user would need to do to update it. For example, the price in each Flavor might be updated by an administrator. Do you have a test that covers that operation?

Look at each relationship and consider how you would create or destroy it. Do you have a test that covers that? For example, what can remove the relationship between an Order and a Flavor? What adds a Flavor to a Catalog?

When you do change those elements, how should the result be reflected in the user interface? For example, removing an item from the Catalog should make it disappear from subsequent catalog browsing. But should it disappear from existing Orders?

Perform or plan tests to exercise all of those changes and to verify the results.

Finally, go through the user actions described in the user stories, storyboards, or requirements document, and think about how the actions affect the relationships that you have drawn. If you find an action that has no effect on the entities in your diagram, add some. For example, if a customer can choose a favorite Flavor, we would have to add a relationship to show which flavor the customer has chosen. Then we can ask, how is that relationship deleted? And do we have tests for creating and deleting it?

States

Think about the states of the types of items that you have identified. What interesting states can you infer from the requirements (not from the code, of course)? Different states are those that make a substantial difference to the actions that users can perform on them, or on the outcomes of actions.

For example, on most sales websites, an order changes from being fairly flexible—you can add or delete things—to being fixed. On some sites, that change happens when payment is authorized. On other sites, the change of state happens when the order is dispatched.

Draw a diagram of states for each type of object that you identified in the entity-relational diagram. Draw as arrows the transitions that are allowed between one state and another, and write next to each transition the actions that cause it to happen.

Order

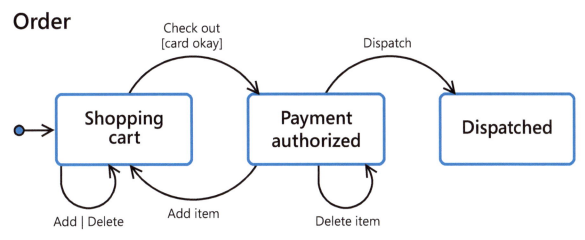

States and transitions

Work out what indicates to the user which state the object is in. For instance, in the system in this example, you can add an item to the order even after you've checked out. There must be some way for the user to be able to tell that they have to check out again.

Devise tests to verify that in each state, invalid operations are disallowed, and that the valid operations have the correct effects. For example, you should not be able to add or delete items from a dispatched order. Deleting an item from an order in the Payment Authorized state should not require you to check out again.

Using models in testing

CRUD and state testing are basic forms of model-based testing.

You can create the diagrams in two ways. One way is to draw them while doing exploratory testing, to help you work out what you are seeing. When you've worked out a model this way, verify that the behavior it represents is really what the users need.

The other way is to draw them in advance to help clarify the ideas behind the user stories when you are discussing them with stakeholders. But don't spend too much time on it! Models have a poor reputation for holding up development, and working code is the most effective vehicle for requirements discussions. Nevertheless, if you can sketch a workflow in a few minutes and have the client say "That's not what I meant," you'll have saved a substantial chunk of work.

MOSTLY NEGATIVE TACTICS

Most programs are vulnerable to unusual sequences of actions or unexpected values. Checking for error conditions is one of the more tedious aspects of computer programming, and so it tends to be skimped on. Such vulnerabilities provide the ideal entry points for hackers. Therefore, it's good practice to push the envelope by producing illogical and unexpected user actions.

Script variation

Work through the steps defined in the test case, but vary the script to see what happens. There are variations that are often productive; for instance, you might:

- Insert additional steps; omit steps. Repeat steps. Re-order major groups of steps—both in ways that ought to work, and ways that should not.
- If there's a Cancel button, use it, and repeat.
- Use alternative ways of achieving the same result of a step or group of steps. For example, use keyboard shortcuts instead of menu items.
- Combine and intermix different scenarios. For example, use one browser to add to an order on a sales website, while paying for the same order with another browser.
- Cut off access to resources such as a server. Do so halfway through an operation.
- Corrupt a configuration file.

The operational envelope

Thinking some more about the state diagrams, consider what the preconditions of a transition between states must be. For example, an order should not be fulfilled if it has no items. Devise tests that attempt to perform actions outside the proper preconditions.

The combinations of states and values that are correct are called the *operational envelope* of the system. The operational envelope can be defined by a (typically quite large) Boolean expression that is called an *invariant*, which should always be true.

You can guess at clauses of the invariant by looking at the entity-relational diagram. Wherever there are properties, there is a valid range of values. Wherever there are relationships, there are valid relationships between properties of the related items. For example:

> Every dispatched Order should always contain at least one item
> AND the total price of every Order must always be the sum of the prices of its items
> AND an item on an Order must also appear in the Catalog
> AND ....

The developers sometimes have to work quite hard to make sure the system stays within its operational envelope when actions occur. For example, if an open order can contain no items, how does the system prevent it from becoming dispatched? For another example, what should happen to outstanding orders when an item is deleted from the catalog? Or is that clause in the invariant wrong?

Devise tests to see if something reasonable happens in each of these cases.

Notice that by trying to express the boundaries of the operational envelope in a fairly precise invariant, we have come across interesting situations that can be verified.

Exploratory and scripted testing in the lifecycle

At the beginning of each iteration, the team and stakeholders choose the requirements that will be developed in that iteration. As a tester, you will work with the client (or client proxy, such as a business analyst) and developers to refine the requirements descriptions. You will create specific test cases from the requirement work items, and work out test steps that should pass when the code has been written. Discuss them with the development team and other stakeholders.

You might need to develop test data or mock external systems that you will use as part of the test rig.

As the development work delivers the first implemented requirements of the iteration, you typically start work on them in exploratory mode. Even if you follow a test case in which you have written steps, you can perform a lot of probing at each step, or go over a sequence of steps several times. (To go back to the beginning of the steps, use the **Reset** button.) Make sure to check the **Create an action recording** option when you start exploration, so that you can log what you have done if you find a bug.

After some exploration, you can be more specific about test steps, and decide on specific values for test inputs. Review your script, adjusting and being more specific about the steps. Add parameters and test values. Record your actions as you go through the script so that it can be played back rapidly in future.

SMOKE TESTS

You don't have to stop the playback at the end of each step. You can if you want to get the test runner to play back the whole test in a single sequence without stopping. A drawback of this is that you don't get a chance to verify the outcome of each step; the whole thing goes by in a whirl. Still, it sure gets the test done quickly, and you can verify the results at the end provided that your last step wasn't to close and delete everything.

A test run played back like this can still fail partway through. If an action doesn't find the button or field that it expects, the run stops at that point. This means that if the system responds in an obviously incorrect way—by popping up a message or by going to a different screen than it should, for instance—then your test will expose that failure. You can log a bug, taking a snapshot of the system state at that point.

This gives us an easy way to create a set of smoke tests; that is, the kind of test that you run immediately after every build just to make sure there are no gross functional defects. It won't uncover subtle errors such as prices adding up to the wrong total, but it will generally fail if anything takes the wrong branch. A good smoke test is an end-to-end scenario that exercises all the principal functions.

You don't have to be a developer to create these kinds of tests. Contoso's Bob Kelly is a tester who has never written a line of code in his life, but he records a suite of smoke tests that he can run very quickly. There's one test case for an end user who orders an ice cream, another for tracking an order, and other test cases for the fulfillment user interface.

Using his smoke tests, Bob can verify the latest build in much less time than it would take him to work through all the basic features manually. When he gets a new build, he usually runs the smoke tests first, and then goes on to perform exploratory tests on the latest features. As he refines those explorations, he gradually turns them into more recorded sequences, which he adds to what he calls his smoke stack.

Monitoring test progress

In this section we will see how we can monitor the progress of our test plan and test suites using Microsoft Test Manager and using reporting in Team Foundation Server.

TRACKING TEST PLAN PROGRESS

While using Microsoft Test Manager, you and your team can do what Fabrikam does—easily monitor the progress of your current test plan and the test suites within them. Your team can also leverage testing reports generated in Team Foundation Server to track your testing progress. These reports are accessible through Team Explorer, Test Dashboard, or Team Web Access, but are also easily shared with other critical people such as business decision makers because they are in Excel format.

Tracking test suite progress in Microsoft Test Manager

In Microsoft Test Manager, you can track your progress for the test suites in your current test plan immediately after you run your tests. You can view the tests that have passed and failed. You can mark tests as blocked or reset tests to active when you are ready to run them again.

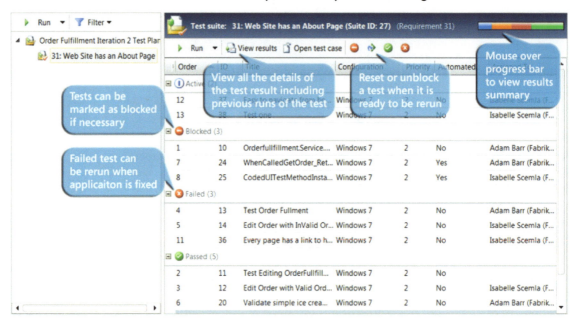

Tracking test suite progress

If you want to view the results for all the suites in the test plan rolled up for your overall status, you can do so in the **Properties** view of your test plan in Test Plan Status.

Tracking test plan results in Microsoft Test Manager

You can also monitor the progress of your test plan by using the test plan results feature in Microsoft Test Manager. The test plan results include charts and numerical statistics on the tests in your test plan. The statistics include the tests that are currently passed, failed, blocked, inconclusive, warning, and active. Additionally, the test plan results include detailed charts that show the failure types and resolution data.

The test plan results can be filtered to specify the test suites and test configurations that you want to be included. For example, you might only want to view the test results for a specific test suite in your test plan that your team is currently active in. Additionally, you might filter the test configurations to only view the test results set to Windows 7. By default, all of the test suites and test configurations that are in your test plan are included in the test plan results.

After you apply the filtering, you can view your test plan progress in either of the two following ways:

- **By Test Suite**, displays the test result statistics for all of the tests in the specified test suites and test configurations in your test plan. This is the default view. It's a quick way to view the progress being made for your test plan. If your test suites are organized by specific iterations, or by particular features or functionality, this can help the team identify areas that are troublesome.
- **By Tester**, displays the test result statistics for all of the tests in the specified test suites and test configurations in your test plan according to which testers performed the tests. This can be useful for load-balancing your tests among your team members.

Test plan results

For more information, see *How to: View Test Plan Results in Microsoft Test Manager.*

Leveraging test reports to track testing progress

In addition to leveraging the tracking information presented in Microsoft Test Manager, you can also track the progress of your team's testing progress using reports. Several predefined test reports are included in Team Foundation Server. You can also create custom reports to help meet a specific reporting need for your team. The predefined reports are available only when your team uses Microsoft Test Manager to create test plans and run tests. The data for the reports is automatically collected as you run tests and save test results. For more information, see the MSDN topic *Reporting on Testing Progress for Test Plans.*

The predefined reports are created for use with Excel. Additionally, a subset of the reports is also available for use with Report Designer. If you create your own tests, you can use them in either Excel or Report Designer.

Your team can leverage the following reports to aide in tracking testing progress in your current cycle.

Tracking how many test cases are ready to run. You can view the progress on how many test cases are ready to run and how many have to be finished for a given timeframe. For more information, see the MSDN topic *Test Case Readiness Excel Report.*

Tracking your test plan progress. You can use the Test Plan Progress report to determine, for a given timeframe, how many test cases were never run, blocked, failed, or passed. For more information, see the MSDN topic *Test Plan Progress Excel Report.*

- Tracking progress on testing user stories: The User Story Test Status report shows how many tests have never run, are blocked, failed, or passed for each user story. For more information, see the MSDN topic *User Story Test Status Excel Report (Agile).*

- Tracking regression: The Failure Analysis report shows the number of distinct configurations for each Test Case that previously passed and are now failing, for the past four weeks. For more information, see the MSDN topic *Failure Analysis Excel Report.*

- Tracking how all runs for all plans are doing: You can use the Test Activity report to see how many test runs for all test cases never ran, were blocked, failed, and passed. For more information, see the MSDN topic *Test Activity Excel Report.*

For more information, see the MSDN topic *Creating, Customizing, and Managing Reports for Visual Studio ALM.*

> **Note:** *There is a delay between the time the test results are saved and the when the data is available in the warehouse database or the analysis services database in Team Foundation Server to generate your reports.*

You can access test reports in one of following three ways:

- Test Dashboard. If your team uses a project portal, you can view the predefined reports on the Test Dashboard. You can access the project portal from the Track view in Microsoft Test Manager. For more information about the Test Dashboard, see the MSDN topic *Test Dashboard (Agile).*

- Team Explorer. You can access Report Designer reports from the Reports folder for your team project, and you can access Excel reports from the Documents folder.

- Team Web Access: If you have access to Team Web Access, just as with Team Explorer, you can access Report Designer reports from the Reports folder for your team project, and you can access Excel reports from the Documents folder.

We're not done till the tests all pass

Don't call a requirement implemented until you've written all the tests that it needs, and they have all been run, and they all pass. Don't call the sprint or iteration done until all the requirements scheduled for delivery have all their tests passing. The chart of requirements against test cases should show all green.

Expect the requirements to change. Update the test cases when they do.

Benefits of system testing with Visual Studio

THE OLD WAY

In a traditional shop like Contoso, system testing simply means installing the product on some suitable boxes, and running it in conditions that are as close as possible to real operation. The test and development teams are separate.

Testers play the role of users, working through a variety of scenarios that cover the required functionality and performance goals.

When a bug is found, they try to find a series of steps to reproduce it reliably. The tester creates a bug report, typing in the steps. The developers try to follow these steps while running under the debugger, and may insert some logging code so they can get more information about what's going on.

Reproducing the bug is an imperfect and expensive process. Sometimes the repro steps are ambiguous or inaccurate. Bugs that appear in operation don't always appear on the development machines.

If the tests are planned in advance, they might be set out in spreadsheets or just in rough notes. If the requirements change, it can take a while to work out which tests are affected and how.

When there's a change to any area of the code, people get nervous. It needs thorough re-testing, and that takes time, and might mean training more people in the product. There's always the worry that a Pass result might be explained by less rigorous testing this time. The biggest expense of making a change—either to fix a bug or to satisfy an updated requirement—is often in the retesting. It's a very labor-intensive endeavor.

THE NEW WAY

Fabrikam uses Visual Studio to improve on this process in several ways:

- **Tests are linked to requirements**. Test cases and requirements (such as user stories, use cases, performance targets, and other product backlog items) are represented by work items in Team Foundation Server, and they are linked. You can easily find out which test cases to review when a requirement changes.

- **Repeatable test steps**. A good way to clarify exactly what is meant by a particular user story is to write it down as a series of steps. You can enter these steps in the test case. When you run the test, you see the steps displayed at the side of the screen. This makes sure the test exercises exactly what you agreed as the requirement. Different people will get the same results from running the test, even if they aren't very familiar with the product.

- **Record/Playback**. The first time you run through a test, you can record what you do. When a new build comes out, you can run the test again and replay your actions at a single click. As well as speeding up repeat tests, this is another feature that makes the outcome less dependent on who's doing the test. (This feature works with most actions in most applications, but some actions such as drawing aren't always recorded completely.)

- **Bug with One Click**. Instead of trying to remember what you did and writing it in a separate bug reporting app, you just click the Bug button. The actions you performed, screenshots, and your comments can all be included automatically in the bug report. If you're working on a virtual environment, a snapshot of that can be stored and referenced in the report. This all makes it much easier for developers to find out what went wrong. No more "no repro."

- **Diagnostic data collection**. System tests can be configured to collect various types of data while the system is running in operational conditions. If you log a bug, the data will be included in the bug report to help developers work out what happened. For example, an IntelliTrace trace can be collected, showing which methods were executed. We'll look at these diagnostic data adapters in more detail in Chapter 6, "A Testing Toolbox."

- **Requirements test chart**. When you open the Reports site from Team Explorer, or the Project Portal site, you can see a report that shows which requirements have passed all their associated tests. As a measure of the project's progress, this is arguably more meaningful than the burn-down chart of remaining development tasks. Powering through the work means nothing unless there's also a steady increase in passing system tests. Burndown might help you feel good, but you can't use it to guarantee good results at the end of the project.

- **Lab environments**. Tests are run on lab environments—usually virtual environments. These are quick to set up and reset to a known state, so there is never any doubt about the starting state of the system. It's also easy to automate the creation of a virtual environment and deploy the system on it. No more hard wiring.

- **Automated system tests**. The most important system tests are automated. The automation can create the environment, build the system, deploy it, and run the tests. A suite of system tests can be repeated at the touch of a button, and the automated part of the test plan can be performed every night. The results appear logged against the requirements in the report charts.

Summary

Tests are directly related to requirements, so stakeholders can see project progress in terms of tests passing for each requirement. When requirements change, you can quickly trace to the affected tests.

You can perform tests in exploratory or scripted mode.

Exploratory tests, and the design of scripted tests, can be based on the idea of different types of tours, and on business models.

Scripted tests help you clarify the requirements, and also make the tests more reliably repeatable.

Scripted tests can be recorded and rapidly replayed. Also, you can use test impact analysis to focus just on those tests that have been affected by recent changes in the code.

The time taken to reproduce bugs can be significantly reduced by tracing the detailed steps taken by the tester, and by placing a snapshot of the environment in the hands of the developer who will fix the bug.

Differences between Visual Studio 2010 and Visual Studio 2012

- **Exploratory Tester**. In Visual Studio 2010, you can record exploratory tests by creating a test case that has just one step, and you can link that to a requirement. During exploratory testing, you can record actions, log bugs, and take screenshots and environment snapshots.

  In Visual Studio 2012, you can associate an *exploratory test session* directly with a requirement (or user story or product backlog item). In Testing Center, choose **Do Exploratory Testing**, select a requirement, and then choose **Explore Work Item**. In addition to the 2010 facilities, you can create test cases during your session.

- **Multi-line test steps**. In Visual Studio 2012, you can write sub-steps in a manual test case step.
- **Windows apps**. Visual Studio 2012 includes specialized features for testing Windows apps.
- **Performance**. Many aspects of test execution and data collection work faster in Visual Studio 2012. For example, the compression of test data has been improved.
- **Compatibility**. Most combinations of 2010 and later versions of the products work together. For example, you can run tests on Team Foundation Server 2010 using Microsoft Test Manager 2012.

Where to go for more information

All links in this book are accessible from the book's online bibliography available on MSDN: *http://msdn.microsoft.com/en-us/library/jj159339.aspx*.

5   Automating System Tests

Manual testing is the best way to find bugs in new code. But the trouble with manual tests is that they are slow, which makes them expensive to rerun. Even with the neat record/playback feature, someone has to sit at the screen and verify the results.

To make matters worse, as time goes on, you accumulate more functionality, so there are more features to test. You only have limited testing resources. Naturally, you only test the features that have most recently been developed. A graph over time in which test cases are listed on the vertical axis would look like this:

Manual testing

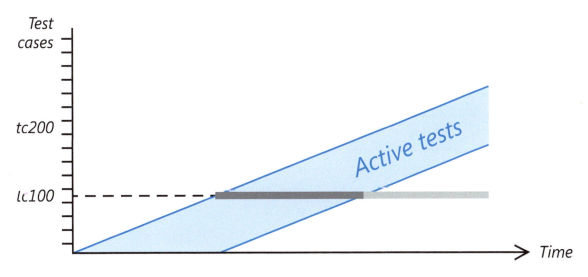

Tests over time

If we follow a particular test case along, it comes into use when the corresponding backlog item is implemented, and maybe gets repeated a few times until the feature is right. The interesting question is, what happens to it after that? Does it just fall into disuse? Or does it get mothballed until just before the big release?

That might have been acceptable for very traditional project plans where they developed each component and then locked it down while moving on to others. But in the past few decades we've discovered that the best way to reduce the risk of a failed project is to plan by user-visible features so that we can get feedback from our clients as we go along. And in any case, even the most traditional projects get revised requirements sooner or later. The downside is that the development of each new user story (or change request) tends to revisit many parts of the code, so that already-developed features might be accidentally disturbed. So we must keep retesting.

Now, we've already said that one of the prime purposes of unit testing is to guard against regressions—to keep the code stable while we work on it. But unit tests aren't quite like system tests. They run on the build server and are not linked to requirements.

This is what automated system tests are for. Like unit tests, they run entirely automatically so that you can run them as often as you like. But they are test cases and can be linked to requirements, so you can always see an up-to-date chart on the project website representing which requirements have all their tests passing. And system tests run on a lab environment, to mimic operational conditions more closely, especially for distributed systems.

This means that on the graph, we can fill that triangle after the manual band with automated testing. The typical test case starts life as a manual test, and then is automated. Okay, you might not automate every test case; but if you can automate a good sample of smoke tests, you will substantially increase your confidence that all the stories that worked yesterday still work today.

WHO CREATES AUTOMATED TESTS, AND WHEN?

Some teams have separate developers and testers. An issue that might concern them is that automated tests obviously involve coding, and not all their testers write much code. If we're suggesting automating the system tests, does that mean retraining all the test staff?

Firstly, let's reiterate that manual tests are the most effective way to find bugs in new code. Automated tests are, on the whole, for regression testing. An experienced tester can nose out a bug much more effectively than test code that simply exercises a predetermined scenario. (Yes, there are techniques like fuzz testing and model-based testing, but if you have the people and tools to do that, you probably don't have many non-coding testers.) So we'll always need manual testing.

Next point: We're about to look at tools that make it very easy to automate a manual test. The learning curve has a shallow end, and the slope is not steep, so you needn't get out of your depth. For the more advanced tests, testers and developers should work together. The tester says what needs to be verified and where.

Lastly, we do recommend that, if your test and development teams are separate, you should mix them up a bit. We have heard of startling improvements in test performance just by introducing one developer into a test team. Similarly, developers can learn a lot by having a tester close at hand. Developers tend to take a kindly view of their code, and don't like to see it badly treated. The tester's attitude to vulnerabilities can have the effect of toughening them up a bit.

Many companies make less distinction between testers and developers than they used to, and some make no formal distinction at all. A team naturally has different strengths, and you'll have some who are better at testing and some better at creating new features. But one skill informs the other, and a mixture of skills is beneficial, especially if it occurs within the same person. Indeed, the best test engineers are expert developers: they understand the architecture and its vulnerabilities. In Microsoft, the job of software development engineer in test (SDET) is well respected, and denotes someone who doesn't simply code, but can also devise sophisticated harnesses that shake the software to its roots. If you have a separate test team that doesn't do development, we recommend introducing at least one developer into the team; we know of cases where doing so has brought substantial improvements.

How to automate a system test

There are several aspects of automating a system test that are, to some extent, independent.

- **Code the test method**. There are two different ways to create the test:
  - **Coded UI test (CUIT)**. You record your actions as you work through a test manually. Then you use the CUIT tools to turn the recording into code.
  - **Write an integration test manually**. You write it exactly as you would a unit test, but instead of aiming to isolate a small piece of the system code, you exercise a complete feature. Normally in this method, you drive the layer just below the user interface.
- **Link the test method to a test case** and thereby to requirements, enabling the result of the test to contribute to the charts of requirements status on the project website. You can do this from the associated automation page of the test case.

- **Set up a test plan for automated tests**. You specify what lab environment to use, which machine to run the tests on, what test data to collect, and you can also set timeouts. If you want to run the same set of tests on different configurations of machines, you can set up different test plans referencing different labs.
- **Define the build workflow**. By contrast with unit tests, which typically run on the build server, a system test runs on a lab environment. To make it run there, we have to set up a workflow definition and some simple scripts.

When you're familiar with all the necessary parts, you might automate a test in the following order:

1. Create a lab environment. For automated tests, this is typically a network-isolated environment, which allows more than one instance of the environment to exist at the same time.

2. Set up a build-deploy-test workflow.

3. Create the code for the tests. You might do this by running the test manually on the lab environment, and then converting it into code.

However, to make things simpler, we'll start by coding the tests, without thinking about the lab environment. Let's assume that we're coding the tests either for a desktop application, or for a web site that is already deployed somewhere. You can therefore do this coding on any convenient computer.

CODE THE TEST METHOD

Prerequisites

On a suitable computer, which can be a virtual lab machine, you must have:

- Visual Studio Professional or Ultimate in order to run Microsoft Test Manager (MTM).
- Visual Studio Premium or Visual Studio Ultimate in order to convert tests into code.
- The client application of the system under test. If you are testing a website, this will just be a web browser.

MTM and Visual Studio can be on different machines, but in that case the client application must be installed on both of them.

Unless the application is stand-alone, the server must be installed somewhere, such as a lab environment.

Generate the coded UI test

You can generate test code from a recording that was made during a manual test case. You can't automate an exploratory test, but you can create a test case with just one step, and record any actions you want in that step.

1. Run the manual test, choosing to create an action recording. Save the test run.

   Play back the recording to make sure it works.

2. In Visual Studio, create a Coded UI Test project in a separate solution from the system under test. You'll find the template under Visual Basic\Test or Visual C#\Test. In the dialog box, choose **Use existing action recording**.

Or, if you already have a CUIT project, open CodedUITest*.cs, right-click anywhere in the code, and choose **Generate Code for Coded UI Test, Use existing action recording**.

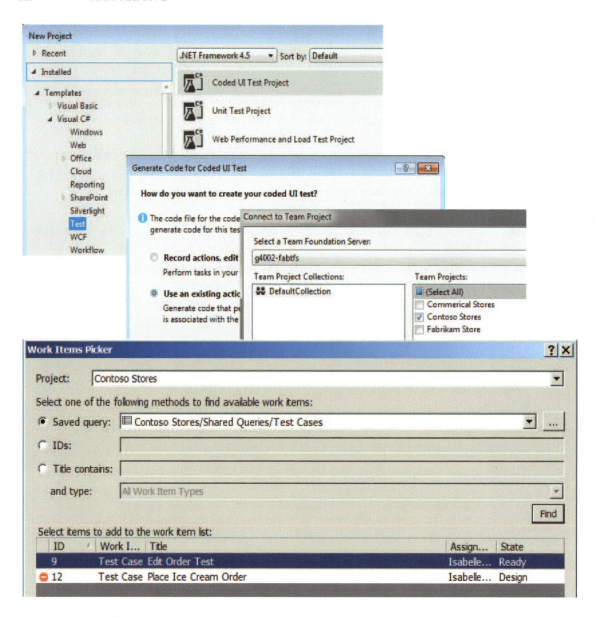

Generating a coded UI test

3. Select the test case in the work item picker.

   New files and code are added to the project. In CodedUITest*.cs you will find code representing the steps of your test:

```C#
[DataSource(...)]
[TestMethod]
public void CodedUITestMethod1()
{
    this.UIMap.LaunchUI();
    this.UIMap.ClickEditOrderfromleftMenu();
    this.UIMap.EnterValidOrderIDOrderId();
    this.UIMap.ClickGetOrderbutton();
    this.UIMap.EditQuantityfrom2to1();
    this.UIMap.ClickUpdateButton();
}
```

Code for the individual steps has been added to UIMap.uitest.

4. On the **Unit Test menu**, choose **Run Unit Test**, **All Unit Tests**.

   Note: *Do not touch the mouse or keyboard while the coded UI test is running. Allow a minute for it to start.*

The test runs just as it did when you played it back in MTM.

Edit the steps

Tip: *Use the UI builder tools to edit your coded UI test where possible. You can use the UIMap Builder to insert new material. Only the top level steps have to be edited in the actual source code. More sophisticated adaptations will require coding, but the basic steps can be created using the tools.*

To rearrange or delete major steps, edit the content of CodedUITest*. These steps correspond to the steps of the test case.

To insert more major steps, place the cursor anywhere between one statement and the next, and on the shortcut menu, choose **Generate Code for Coded UI Test, Use coded UI builder**. The UIMap builder appears on at the bottom right of your screen:

UIMap builder

Start your application and get it to the state you want before your new actions. Then press the record button (the one at the left of UIMap builder) and perform your new actions. When you're finished, press Pause, and then Generate Code (the button at the right). Close the UIMap builder, and the new code will appear in Visual Studio where you left the cursor.

To edit the actions inside any step, open UIMap.uitest. This file opens in a specialized editor, where you can delete or insert actions. These are the methods called from the main procedure. Notice that they are in no particular order.

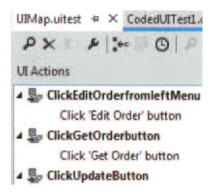

Edit actions

> **Note:** *Do not directly edit the code in UIMap.Designer.cs—your edits will be overwritten. Instead, use the CUIT Editor by opening UIMap.uitest; or write your methods in UIMap.cs, which will not be overwritten. There is a menu item for moving code into UIMap.cs.*

To make a step optional—that is, to ignore the error that would normally be raised if the UI element cannot be found—select the step in UIMap.uitest, and then set the **Continue on error** property.

Validate values

In CodedUITest*.cs, place the cursor at the point where you want to verify a value. On the shortcut menu, choose **Generate Code for Coded UI Test, Use existing action recording, use coded UI builder.**

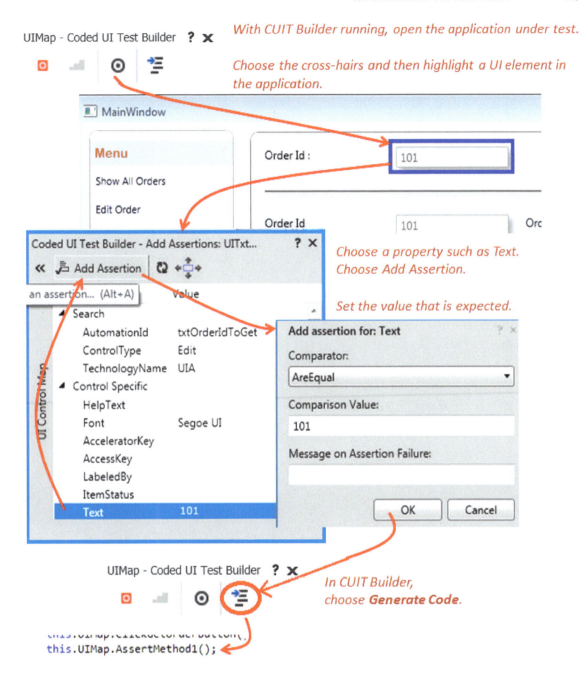

With CUIT Builder running, open the application under test.

Choose the cross-hairs and then highlight a UI element in the application.

Choose a property such as Text. Choose Add Assertion.

Set the value that is expected.

In CUIT Builder, choose **Generate Code**.

Validating test results

Open the application under test.

Drag from the crosshairs in the CUIT Builder to any field in the application window. The CUIT Properties window opens, where you can choose a property—typically **Text**. Choose **Add Assertion** and specify the value that it should take. Then in CUIT Builder choose **Generate Code**. A method is added to UIMap, and a call to it is added to your test code.

Also notice that AutomationId is set and can be used to search for the UI Element to be used in the CUIT.

> **Tip:** *Don't choose any of the Search properties.*

Notice that you don't need to be very skilled in development to create a basic test that includes value validations.

Data-driven tests

You can make a test loop multiple times with different data. The easiest way to do this is to insert parameters in your test case before you record a run. Parameters are the feature of manual tests in which you can write a test step such as "Select a @flavor ice cream of size @size." While editing the test case, you fill in a table of values for your parameters, and when you run the test, Test Runner takes you through the test multiple times, once for each row of parameter values.

When you generate code from a parameterized test case, the code includes the parameter names.

You can later change the values in the parameter table. When you play back the actions, or when you run the code as an automated test, the new values will be used.

> **Note:** *Before you record the actions for a parameterized test, just provide one row of parameter values so that you will only have to work through the test once. When you have completed the manual run, generate code from it. Then write more values in the parameter table for the automated test.*

If you would rather supply the test data in a spreadsheet, XML file, or database, you can edit the DataSource attribute that appears in front of your test. It is initially set to gather data from your parameter list:

```C#
[DataSource("Microsoft.VisualStudio.TestTools.DataSource.TestCase",
            "http://g4002-fabtfs:8080/tfs/defaultcollection;Commercial Stores",
            "12",
            DataAccessMethod.Sequential)]
[TestMethod]
public void CodedUITestMethod1()
{...
```

However, you can change it to use any other data source that you choose. For example, this attribute gets data from a comma-separated value (CSV) file:

```C#
[DataSource("Microsoft.VisualStudio.TestTools.DataSource.CSV",
"|DataDirectory|\\data.csv",
            "data#csv", DataAccessMethod.Sequential),
DeploymentItem("data.csv")]
[TestMethod]
```

The first line of the CSV file should be the comma-separated names of the parameters. Each following line should be comma-separated values for each parameter. Add the file to your project, and set the **Copy Always** property of the file in Solution Explorer.

Further extensions

CUITs provide a very easy entry point for creating automated tests. However, there are several ways in which you can write code to adapt and extend the basic recording. You also have to write code to handle more complex user interfaces.

If you add methods to the project, write them in separate files. Don't modify UIMap.Designer.cs, because this file is regenerated when you use the UI builder, and your changes will be lost. If you want to edit a method that you initially created using the UI builder, move the method to UIMap.cs and edit it there.

Here are some of the things you can do with hand-written code. The details are on MSDN under *Testing the User Interface*:

- *Data-driven tests*. Loop through your test multiple times with data taken from a spreadsheet or database.
- *Write code to drive the interface and keyboard*. The UIMap provides easy-to-use functions such as Mouse.Click(), Keyboard.Send(), GetProperty().
- *Wait for events* such as controls appearing or taking on specified values. This is useful if the application has background threads that can update the user interface.
- Extend the CUIT recorder to interpret gestures for new UI elements.

Bob, Lars is going to help you automate some of your tests.

I knew this was coming. I don't write code. I just test stuff.

Visual Studio! I only use MTM...

Hi! So I'll just fire up Visual Studio.

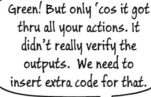

Run a manual test with Recording switched on. Scripted or exploratory, doesn't matter. Then drop by and we'll code it together.

Now click "Generate coded test"... "From action recording"... There you go! Code. Now CTRL+R,A and let's see it run...

Green! But only 'cos it got thru all your actions. It didn't really verify the outputs. We need to insert extra code for that.

So I'll take over. Tell me all the points where it should check a value, and I'll create the asserts. They're not that hard, really.

That's really cool! Next time, I'll turn it into code myself — if I can get you to help with the assertions...?

Sure!
Main thing to look out for tho' is changes in the UI. Layout changes are OK, but if there's a change in sequence, you need to do some work with the UIMap editor.

But you know, I think after you've seen me do it a couple of times, you'll be inserting the asserts yourself. The CUIT Builder helps you create them, so there's no big programming. You can always come to me for help with complex ones.
We'll have heaps of your tests coded up in a few weeks!

Using coded UI tests

Tips for using CUITs:

- Don't touch the keyboard or mouse while a CUIT is playing.
- Use the CUIT Builder and Coded UI Test editor, rather than updating the code directly.
- Create separate methods for separate dialogs.
- Use the Split function in the CUIT editor to refactor a long method into two.

> **Note:** *For CUITs to work effectively and to be robust against changes, each UI element has to have a unique Automation ID. In HTML, each interaction element must have an ID. Then, you can change the layout and appearance of the application and the CUITs will still work.*

Coding integration tests by hand

An alternative way to automate an existing manual test is to write by hand some code that works through the same sequence as the manual steps. You can then associate it to the test case.

Typically, you would write such a test as an integration test. Instead of driving the user interface, your test would use the business logic that is normally driven by the UI.

You create these tests exactly as you would a unit test. If there is a public API to the business logic, create the test in a solution separate from the system under test.

Advantages and disadvantages of integration tests are that:

- They work in cases where you can't record a CUIT.
- They are more robust than CUITs against changes in the UI.
- They require a clean separation between business logic and UI, which is good.
- They don't test any validation or constraint logic that the UI might contain, and which gives the user immediate feedback.
  You could therefore miss a bug that results from the UI allowing an input that the business logic doesn't expect. This is perhaps the strongest reason for preferring CUITs where they are feasible.
- It takes longer to code a test than to record the actions of a CUIT. However, since you're likely to want to do some work on the code of the CUIT to generalize it, you might find that this difference becomes less significant.
- There is no guarantee that the method you write tests the same thing as the corresponding manual test steps (if there are any). You could create a test case that fails when performed manually and passes when executed automatically.

LINK THE TEST METHOD TO THE TEST CASE

Now that you have a working test method, you can promote it to be the test method that automates a test case. If you derived the test method from the steps of a recorded test case, then that is the test case to link to.

Test cases are usually linked to requirements. Linking a test method to a test case allows the results of the test to contribute to the report of test results for each requirement.

1. Check in the test method.

   Open the solution containing the test method, if it is not already open.

2. Link the test case with the test method:

   Open the test case in Team Explorer. On the **Associated Automation** tab, choose the ellipsis button and select the test method. The Automation status will change to Automated.

Automation status **Automated**

STEPS	SUMMARY	TESTED USER STORIES	ALL LINKS	ATTACHMENTS	**ASSOCIATED AUTOMATION**

Automated test name

Orderfullfilment.Automation.EditOrder.EditOrderTest

Automated test storage

orderfullfilment.automation.dll

Automated test type

CodedUITest

Remove association

3. In Microsoft Test Manager, make sure that the test case appears in a test suite, such as a requirement-based test suite.

To create linked test cases for a batch of test methods, there is a command-line tool available on MSDN, *tcm.exe*.

CREATE AN ENVIRONMENT FOR AUTOMATED TESTS

When you set up a build workflow, you will specify a particular lab environment on which it will run.

We discussed how to set up a lab environment in Chapter 3,"Lab Environments," but there are some specific points you need to know for automated tests.

- Use an SCVMM environment so that you can take a snapshot and revert to it before starting each test.
- You need a machine for each component of your system, including the client machines. For example, the Fabrikam Ice Cream website has: a client machine with a web browser; a machine for the web server on which Windows 2008 is installed with Internet Information Services (IIS) enabled; a database server; and a client machine for order fulfillment.
- In the New Environment wizard, on the **Advanced** tab, **check Configure environment to run UI Tests** and choose the client machine. This makes the test agent run as a desktop application. Before tests start, it logs into the machine with the credentials you supply.
- Take a snapshot of the environment. Note its name: in the build definition; you will specify that the tests should start by reverting to this snapshot.

- You have to designate this environment as the one on which your test plan will run. Select the environment in **Test Plan > Properties > Automated Test Settings**.

Network isolated environments

If you will be running several tests of the same type, it is worth a bit of extra effort to set up a network isolated environment (NIE). An NIE allows you to deploy more than one copy of it from the library without experiencing naming conflicts. The NIE has its own internal private network: the machines can see each other, but they cannot be seen from the external network.

Another advantage is that pre-installed software is not affected by deployment because the machine identities are not changed when they are deployed. This avoids having to reconfigure software before using a copy of the environment.

Lastly, bugs can be easier to trace because the machine names are always the same.

You have to do a little bit of extra work to set up network isolation. We'll summarize here, and you can find more details in the MSDN topic How to: *Create and Use a Network Isolated Environment*.

The additional points to note are:

- You need an additional server machine that you configure as a domain name server. Log into it and enable the Domain Controller role of Windows Server.
- Set the Enable Network Isolation option in the environment wizard.
- When you have started the environment, log into each machine and join each machine to the private domain provided by the domain controller.
- Stop the environment and store it in the library.

SET A TEST PLAN TO PERFORM AUTOMATED TESTS

Set up one or more test plans to execute the tests. The test plan determines which lab environment will be used.

In Microsoft Test Manager, choose Plan, Properties, and then under Automated runs, choose a test environment.

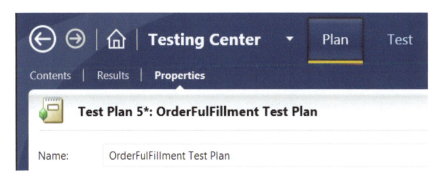

The test settings selection within the test plan properties

You can usually leave Test Settings at Default. You can use different settings to: filter the list of environments to those with a specific set of roles; vary the data collected from each machine; have additional data and scripts downloaded to the test machines before the tests begin; set timeouts.

AUTOMATED DEPLOYMENT TO A LAB ENVIRONMENT

It's good to install your system the same way the users will, to make sure their experience is satisfactory. If your system has different components on different machines, then you'll typically ask them to log in to each machine and run an installer there.

To get your installation fully automated, you need to get two things working properly: the creation of the setup files when the source code is built and the running of the setup file on the target machine in the lab environment.

When you set up a build-deploy-test workflow, the test controller and the test agents in the lab machines will do one part of the setup task for you. On each lab machine, it will run a script that you specify. Your script then has to do whatever is necessary to copy the setup files onto that machine, and run the setup there.

In Visual Studio 2012, there are several deployment mechanisms. The principal ones are:

- **Windows Installer**. To create the familiar setup and .msi files, add a setup project to your solution. The project template is under Other Project Types. In the project properties, specify which of the solution's projects and files you want to include in the installation. When you build the project—either on the development machine or in the build server—the setup files are created in the project's output folder.

  If your system has components that are installed on different machines, then you will create one setup project for each component. Each installer can include assemblies built in more than one project.

- **ClickOnce** deployment makes it particularly easy to deploy updates to a client or desktop application. You click one button in Visual Studio to deploy it to a specified public or intranet location. Every time the user starts the application, it looks in that location for updates.

  ClickOnce applications have some limitations. There are additional security checks to allow them to access files. The dialogs in the update mechanism make ClickOnce more appropriate for user than service software. Your application has to be defined in a single project.

- **Website publication** allows you to upload a website to a web server by using an agent that you can add into IIS. You can also upload a database at the same time.

See the MSDN topic *Choosing a Deployment Strategy* for a detailed comparison.

Let's consider how to configure an automated build that uses the first two mechanisms. (Websites can be installed using a setup project, so we won't consider website publication.)

Automating deployment with Windows Installer

The lab build workflow

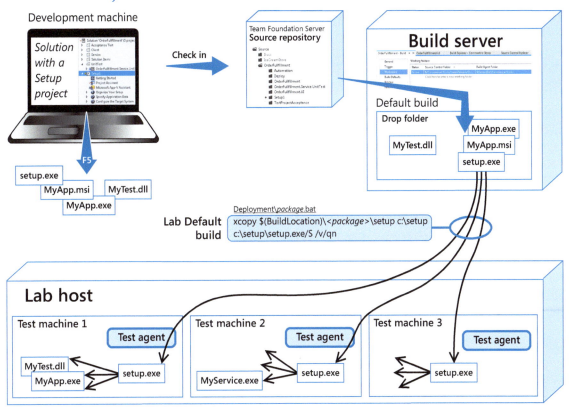

Lab deployment using Windows Installer

The diagram shows the flow of code from the developer's machine to the lab machines in the lab environment.

The solution includes a setup project, which generates an .msi file and setup.exe. The setup project is checked in along with the test code and application code. Any build on the build server creates the same files in the same relative locations, but in a drop folder.

The lab build then goes on to install the assemblies on the test machines and runs the tests there. This is a step beyond the kind of build we met in Chapter 2, "Unit Testing: Testing the Inside," where the tests usually ran on the build server.

Writing a deployment script

When you define the lab build, you will specify scripts that deploy the installers on the lab machines. You need a separate script for each lab machine. It's best to write and debug the scripts before you create the build definition.

Here is a typical script:

```
rem %1 is the build location – for example \\buildMachine\drops\latest.
mkdir c:\setup
rem Replace OrderFullfillmentSetup for each test machine:
xcopy /E /Q %1\OrderFullfillmentSetup\* \* c:\setup
cd c:\setup
rem Install quietly:
setup.exe /S /v/qn
```

The script expects one parameter, which is the UNC path of the drop folder; that is, the location on the build server where the setup file will be deposited. This particular script will get the setup files from the OrderFullfillmentSetup project; each test machine should look in the subfolder for its corresponding setup project.

The parameters to setup.exe on the last line encourage it not to pop up dialogs that would prevent the installation from continuing.

Test your deployment script. Run a server build manually, and then log in to one of your test machines. Copy the script there, and call it, providing the build drop location as a parameter.

Don't forget you need a separate script for each machine.

Place your deployment scripts in your Visual Studio solution. Create a new project called Deploy. Add the deployment scripts to this project as text files. Make sure that each file has the **Copy To Output Directory** property set to Copy Always.

Lab and server builds

To compile and run checked-in code, you create a build definition. You can see your project's build definitions under the Builds node in Team Explorer. There you can run a build manually, and open a report of the results of its most recent run.

There are two main kinds of build. Build definitions are created from build definition templates, and there are two template provided out of the box. The template that you get by default is helpfully named the Default template, and when you define a build with it, the tests usually run on the build server. We will call this kind a server build:

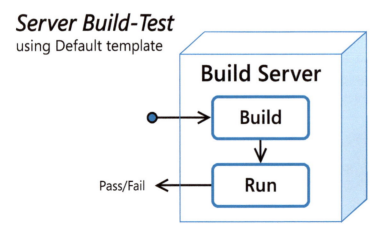

Server build runs tests on the build machine

The other kind of build is created from the Lab Default template. In this kind of build, the build server is used only to compile the source code. If the compilation is successful, the components are then installed and run on machines in a lab environment. Deployment on each machine is controlled by scripts that you write. This enables the complete build-deploy-test scenario.

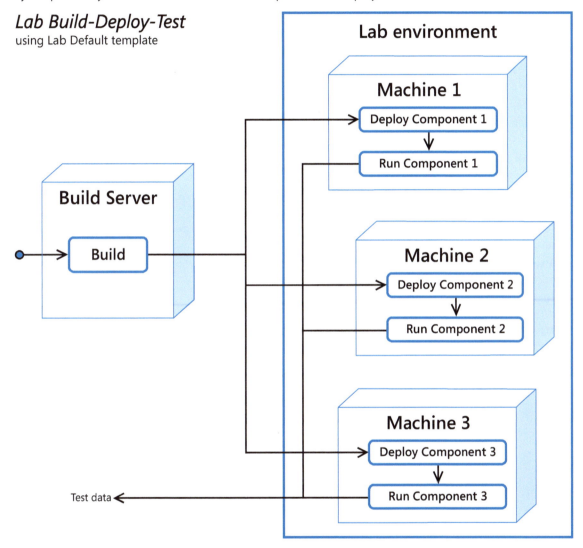

Lab build definition runs tests on lab machines

Builds defined with the Lab Default template are more like workflows. They do the following:

1. Build the code for the application, tests, and setup on the build server. To do this, the build invokes a default build definition.

2. Execute a script of your choice on each test machine in the lab. You would normally provide a script that copies the setup files to the test machine and executes them there. (But you can provide a script that does anything.)

3. Invoke tests on the lab machine that you designate as the test client. They run in the same way as unit tests.

What's in a build definition?

Lab and server build definitions are MSBuild workflows. They have several properties in common, such as the trigger event that starts the build. To define a lab definition, you also specify:

- The **lab environment** on which to deploy the system. Typically you would specify a virtual environment, so that if a test fails, a snapshot can be logged.

- **Deployment scripts** that perform the deployment. You have to write these scripts. We'll show you how shortly.

- **Test settings**, which specify what data to collect from each machine. These override the test settings defined in the properties of the test plan.

- A **server build**, which is used to determine what source code to compile. This means that you have to define a server build before you define a lab build.

- **Test suites**, which contain the test cases that will be run. Test results will be reported in terms of test cases and requirements. (By contrast, server builds specify simply the assemblies from which test methods are to be executed, and cannot be related to requirements. The test assemblies specified in the server build definition are ignored by the lab build.)

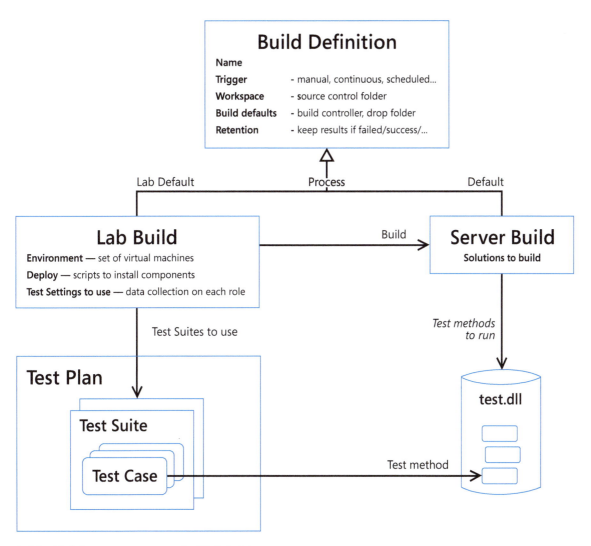

Lab build definition uses a server build definition

Identifying a server build definition

A lab build uses a server build to do its compilation; so before you define a lab build, you must first have a server build defined.

Now you've almost certainly already defined a server build, because you used it to run unit tests on the build server. In fact, you can use one of those definitions. Any server build is suitable, so long as it compiles all the source of the application and tests.

It doesn't matter what unit tests it runs, because they will not be run when it is pressed into service in the lab build. Neither does it matter what its trigger is, because the lab build will start it.

Referencing a server build from a lab build will not prevent it from running according to its own trigger.

But if you would prefer to define a separate server build to be part of the lab build, set its trigger to manual. Run it to make sure that it builds correctly. Refer back to Chapter 2, "Unit Testing: Testing the Inside," for the details.

Creating a lab build-deploy-test definition

Here are the steps you use to create a lab build definition.

1. In Team Explorer, create a new build definition. Actually, it will turn out to be a workflow, rather than a simple build definition.

   Select the trigger condition, such as **Continuous Integration**.

   Select **LabDefaultTemplate**. This means we're creating a lab definition.

   Click Lab Process Settings to open the Lab Workflow properties wizard.

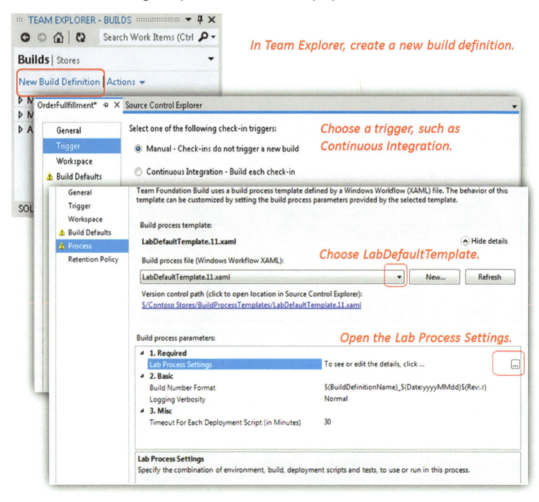

2. In the lab process settings wizard:

In the Environment page, specify that you want the build to start by reverting to your baseline snapshot.

Select the build that defines which source to compile. In the dropdown, the options you see are build definitions that were created from DefaultTemplate. (The build specified in the test plan properties is ignored.)

Add invocations of the deployment scripts that you prepared earlier. Each script is executed on the test machine that you designate.

Set the working directory (unless you make a habit of starting with a change directory command (cd) in the script). Each script is copied to the lab machine and runs there.

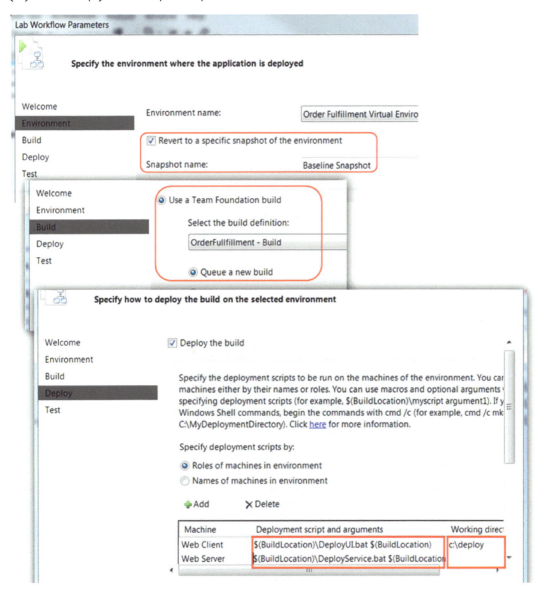

3. In the Test page, specify the test plan and suites that you want to be executed. The test plan determines what lab environment will be used, what test data will be collected, and might also specify additional files to be deployed on the lab machines. The test suites determine which test cases will be run.

Pick a test plan and select test suites.

Select test settings. This overrides the test settings defined in the test plan.

Save the lab build definition.

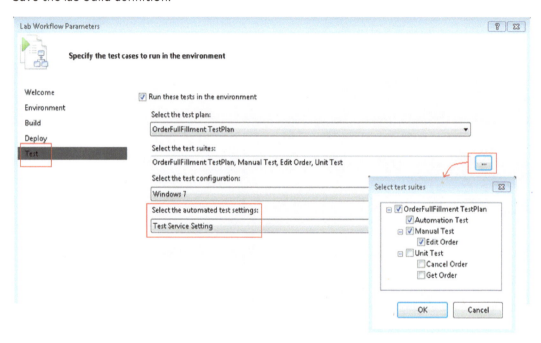

The build will run on the defined trigger.

Automating deployment of ClickOnce applications

ClickOnce applications have a rather different deployment path. This feature allows you to deploy a desktop application straight from Visual Studio by using the Publish Now button in the project properties. There are certain limitations: for example, the application has to have user attention, so it isn't useful for services.

Apart from being easy to deploy from the developer's point of view, the really interesting thing is that whenever a ClickOnce application is started by its user, it checks back to its deployment site to see if there is a new version.

You can put this feature to work in your tests—and test it at the same time.

There are two approaches. One is to install the application on a test machine and allow it to update itself automatically from the latest build. Alternatively, you can use the coded UI test to run the installer explicitly.

Letting the application update itself:

1. When you first set up your test environment, install the application on the appropriate machine. Do this *before* you take the snapshot to which the environment will revert at the start of every test.

2. Add to your server build definition a command script that runs the ClickOnce publication as part of the server build. This is the server equivalent of the developer pressing the Publish button in Visual Studio. To see how to do it, read *Building ClickOnce applications from the command line*.

3. Use an empty deployment script for the test machine on which the application will run.

4. Before you record your test, make sure that a fresh version of the application has been built. Start the recording before you start the application. The application will detect a new version and pop up a dialog asking you for permission to install it. Choose **OK**, acknowledging that the new version is now part of your test.

5. Stop recording, generate the code of the test, and then close UI Builder.

This test will run the application and allow the latest version to install itself.

However, it will fail if you run twice for the same version, because it expects to see the dialog that asks permission to install a new version. To allow your test to work in both situations (with or without a new build), open UIMap.uitest, select that step, and in the Properties window, set **Continue on error** to true.

Using a coded UI test to run the installer

It's a good idea to test the installer explicitly, especially if it has options. You can create a coded UI test to do this.

You will need:

- An environment where the machine on which you will run the installer is set as the machine on which coded UI tests will be run. If you're testing the installer for a desktop application, you already have that. If you are testing the installation of a server, you will need to create a new environment in which a server machine is designated as the coded UI test machine. You do this in the Advanced page of the New Environment wizard.

- A deployment script that copies the installer to the test machine, but does not run it:

```
rem %1 is the build location - for example \\buildMachine\drops\latest.
mkdir c:\setup
rem Replace OrderFullfillmentSetup for each test machine:
xcopy /E /Q %1\OrderFullfillmentSetup\* \* c:\setup
```

- Visual Studio must also be installed on this test machine.

Log in to the test machine, and use Visual Studio to record a sequence in which you run the installer.

You should also run the application, to make sure that it has been installed correctly.

Generate code from the recording.

One of the challenges in dealing with automated tests for installation, updates or removal is that it's common for various messages to occur. Often it's a security issue dealing with security credentials or something else that is occurring in the environment under development.

Again, we could set **Continue on error** for a step where something indeterminate occurs, but it can be useful to log the cause of the error while skipping over it. For this, an explicit try/catch can be useful. Move the code into UIMap.cs so that you can edit it.

Viewing the test results and logging bugs

To display a list of recent test runs, in Microsoft Test Manager, choose **Testing Center, Test, Analyze Test Runs**.

You can edit the title and comment and, if necessary, the reason for failure of any test run.

You can open the details of any individual test and inspect the data collected from the test.

If necessary, you can also create a bug work item, which will automatically include the test data.

Driving tests with multiple clients

Most server-based systems have interesting end-to-end tests that involve more than one client machine. For example, when a customer orders an ice cream through the public web interface, the order should appear on the warehouse interface until it is dispatched. To test that story, we have to write a test that drives first one interface and then the other.

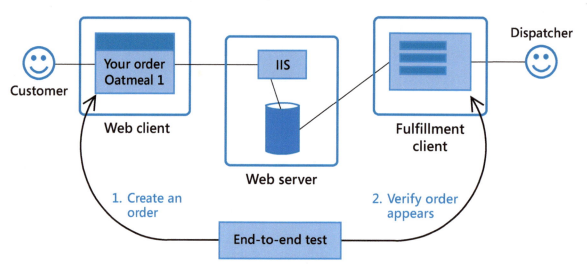

An end-to-end test with multiple client machines

The interesting question is, "Where should we execute the end-to-end tests?" The lab framework assumes that there is a single client machine on which the tests will be run.

Potential schemes include:

- Install all the user interfaces on the same machine. Perform manual tests by driving the user interfaces side by side. From these tests, create coded UI tests that can run on the same configuration.

- Keep the user interfaces on separate machines. Write a proxy for the end-to-end test that you install on each machine. Install the end-to-end controller on the designated client machine, which could be a separate machine, or on one of the user interface machines. To write the proxies, you would generate CUITs and then edit the source to perform actions in response to messages from the end-to-end controller.

You can use a similar strategy to simulate the effect of external systems.

Summary

Automated system tests improve the stability of your code while reducing the costs of running the tests. At the same time, the cost of writing the tests has to be balanced against the benefits. Coded UI tests are a quick and reliable way of creating automated system tests, but must be used with care because they are vulnerable to changes in the user interface.

In this chapter we've seen how to define a continuous build-deploy-test workflow that runs coded tests in a lab environment.

Differences between Visual Studio 2010 and Visual Studio 2012

- **Single test agent**. In Visual Studio 2010 when you prepare a virtual machine for use in automated tests, you have to install the Test Agent, Lab Agent, and Build Agent. These act as proxies for the Test and Build Controllers, installing software, invoking test methods, and collecting test data.
  In Visual Studio 2012, there is just a single agent, the Test Agent. You can install it manually to prepare a virtual machine (VM) for the store. Or, you can have Lab Center install it by using the **Repair** command on an environment.

- **Specialized Test Projects**. In Visual Studio 2010, there is a single type of test project, to which you can add different kinds of test files such as coded UI tests, load tests, and so on. In Visual Studio 2012, there are different types of test projects, to which different combinations of test files can be added.

- **Compatibility**. You can use a combination of 2010 and 2012 version products, and most things work. For example, tests created on Visual Studio 2012 will run in a lab set up in Team Foundation Server 2010.

Where to go for more information

All links in this book are accessible from the book's online bibliography available on MSDN: *http://msdn.microsoft.com/en-us/library/jj159339.aspx*.

6   A Testing Toolbox

It used to be that when you wanted to track down a bug, you'd insert print statements to see where it went. At Contoso, it's still common practice. But Fabrikam's project teams know that there is a wide variety of powerful tools that they can bring to bear not only to discover and track down bugs, but also to keep them to a minimum.

We've seen how Fabrikam's team places strong emphasis on unit tests to keep the code stable, and how they record manual tests to help track down bugs. We've seen how they create virtual environments to make lab configuration faster and more reliable. They like to automate system tests where they can, so as to reserve manual testing effort for exploring and validating new features. They generate charts of tests against each requirement, so that progress (or delay) is constantly evident.

And there are more tools in that toolbox. Let's open it up and see what's in there.

WHERE DO MY TEAM MEMBERS FIND THEIR BUGS?

Using Visual Studio, Microsoft Test Manager, and a few key reports from your build-deploy-test automation, you can easily identify issues in your application under test. Using these Microsoft tools helps locate and resolve bugs in a systematic fashion, thus shortening the cycle time.

Bugs are found by running tests. There are four principal ways in which bugs are found and managed:

- **Local unit testing on the development machine**. Before checking in updated application code, the developer will typically write and run unit tests. In Chapter 3, "Lab Environments," we explained how you can set up check-in policies to insist that specified unit tests are run.
- **Build verification tests**. As the team members check in code to the source repository, the build server runs unit tests on the integrated code. Even if local tests pass, it is still possible for the combined changes from different developers to cause a failure.
- **Manual testing**. If you discover a bug while running a scripted test case or during exploratory testing, you typically generate a bug work item. Alternatively, there might be an error in the test case script.
- **Automated system tests**. After a failed build-deploy-test run, you can inspect the test data, and then either create a bug work item or fix the test.

WHAT OTHER MICROSOFT TOOLS CAN I USE TO FIND BUGS?

After you have your test plan and infrastructure established, you might ask how else Visual Studio can help you find bugs and issues in your applications. That was the question that occurred to the Contoso team after they learned from their new Fabrikam teammates just how many types of tests can be run with the help of Visual Studio, Microsoft Test Manager, and Team Foundation Server. Let's take a look at some other exciting options.

Performance and stress tests in Visual Studio

Visual Studio Ultimate includes performance and stress testing tools, including tools for web performance testing and load testing. Both of these test types can be added to the web performance and load test project in your solution. For more information, see the topic *Testing Application Performance and Stress* on MSDN.

> **Note:** *You must have Visual Studio Ultimate in order to use web performance or load testing.*

Web performance tests: Web performance tests are used to simulate how end users interact with your web application. You can easily create web performance tests by recording the HTTP requests while using your browser session in conjunction with the Web Performance Test Recorder included in Visual Studio Ultimate. You can also create web performance tests manually by using the Web Performance Test Editor included in Visual Studio Ultimate.

Load tests: Load tests provide you with the ability to stress test your application by emulating large numbers of machines and users hammering away at your application. Load tests can include a test mix consisting of any of the web performance tests, unit tests, or coded UI tests that are in the test project of your solution.

In this chapter, we'll also talk about using load tests in conjunction with the automated tests to help discover additional bugs caused by stressing your application.

Adding diagnostic data adapters to test settings

As we discussed in Chapter 4, "Manual System Tests," Microsoft Test Manager provides the testers on your team with the ability to conveniently author, manage, and execute manual and automated tests using a test plan. Let's look at the test settings a bit more thoroughly and see how we can leverage additional functionality to help find and isolate bugs in our application.

Microsoft Test Manager provides your testers with the ability to add and configure additional diagnostic data adapters (DDA), which are used to collect various types of data in the background while tests are running. Diagnostic data adapters are configured in the test settings associated with either Microsoft Test Manager or Visual Studio. When Fabrikam testers saw that Contoso was using different hardware to test network connectivity for LAN, WAN, and dial-up connections, they explained their own process—using a diagnostic data adapter to emulate different kinds of network connectivity, which saves them a good deal of pain.

There are several valuable types of diagnostic data that the diagnostic data adapters can collect on the test machine. For example, a diagnostic data adapter might create an action recording, an action log, or a video recording, or collect system information. Additionally, diagnostic data adapters can be used to simulate potential bottlenecks on the test machine. For example, you can emulate a slow network to impose a bottleneck on the system. Another important reason to use additional diagnostic data adapters is that many of them can help isolate non-reproducible bugs. For more information, see the MSDN topic *Setting up Machines and Collecting Diagnostic Information Using Test Settings*.

Can I create my own diagnostic data adapters?

As Fabrikam happily shared with Contoso team members, you can create your own custom diagnostic data adapters to fulfill your team's particular testing requirements. You can create custom diagnostic data adapters to either collect data when you run a test, or to impact the machine with bottlenecks as part of your test.

For example, you might want to collect log files that are created by your application under test and attach them to your test results, or you might want to run your tests when there is limited disk space left on your computer. Using APIs provided within Visual Studio, you can write code to perform tasks at specific points in your test run. For example, you can perform tasks when a test run starts, before and after each individual test is run, and when the test run finishes. Once again, Contoso saved time and effort by adopting Fabrikam's practices and creating their own data adapters. For more information, see the MSDN topic *Creating a Diagnostic Data Adapter to Collect Custom Data or Affect a Test Machine*.

Test settings for Visual Studio solutions

Test settings for Visual Studio are stored in a .testsettings file that is part of your solution. The Visual Studio .testsettings file allows the user to control test runs by defining the following test roles:

- The set of roles that are required for your application under test.
- The role to use to run your tests.
- The diagnostic data adapters to use for each role.

To specify the diagnostic data to collect for your unit tests and coded UI tests in the test projects of your Visual Studio solution, you can edit an existing test settings file or create a new one. Creating, editing, and setting the active test setting file are all done using the **Test** menu in Visual Studio. To view the steps required to create a new test setting in Visual Studio, see the MSDN topic *Create Test Settings to Run Automated Tests from Visual Studio*. To view the steps used to edit an existing test setting in Visual Studio, see the MSDN topic *How to: Edit a Test Settings File from Microsoft Visual Studio*.

In Visual Studio, when you create a new test project, by default the Local.testsettings is selected for your project. The Local.testsettings file does not have any data or diagnostic adapters configured. You can edit this file, create a new test settings file, or select the TraceAndTestImpact.testsettings file, which has the ASP.NET Client Proxy for IntelliTrace and Test Impact, the IntelliTrace feature of Visual Studio, System Information, and Test Impact diagnostic data adapters.

Test settings for Microsoft Test Manager test plans

Test settings define the following parameters for your test plan in Microsoft Test Manager:

- The type of tests that you will run (manual or automated).
- The set of roles that is required for your application under test.
- The role to use to run your tests.
- The diagnostic data adapters to use for each role.

To view the steps used to create a test setting in Microsoft Test Manager, see the MSDN topic *Create Test Settings for Automated Tests as Part of a Test Plan*.

In Microsoft Test Manager, you can edit existing test settings, or create new ones. The test setting is associated with your test plan, and is configurable in the **Properties** pane. There are separate test settings affiliated with your manual test runs and your automated test runs. You can select the Local Test Run, which by default includes the Actions, ASP.NET Client for IntelliTrace and Test Impact, System Information, and Test Impact diagnostic data adapters.

What about bugs that are hard to reproduce?

An issue that has plagued teams for ages is the non-reproducible bug. We're all too familiar with the scenario in which we find a bug and submit it to a developer only to have the developer come back and say, "I can't reproduce the issue." Sometimes, the bug can go through numerous iterations between the tester or team member who found the bug and the developer attempting to reproduce it. Using either Visual Studio, or Microsoft Test Manager, you can configure your test settings to use specific diagnostic data adapters. For example, the diagnostic data adapter for IntelliTrace is used to collect specific diagnostic trace information to help isolate bugs that are difficult to reproduce. This adapter creates an IntelliTrace file that has an extension of .iTrace that contains this information. When a test fails, you can create a bug. The IntelliTrace file that is saved with the test results is automatically linked to this bug. The data that is collected in the IntelliTrace file increases debugging productivity by reducing the time that is required to reproduce and diagnose an error in the code. From this IntelliTrace file the local session can be simulated on another computer; this reduces the possibility of a bug being non-reproducible. In this chapter, we'll learn more about using IntelliTrace to isolate non-reproducible bugs.

In addition to the IntelliTrace data and diagnostic adapter, you can leverage other adapters, which can help your team find bugs that might otherwise be more difficult to reproduce. For example, you could add the video recorder adapter to help clarify some elaborate steps that are required in order to make an issue occur. You can edit your test settings file, or create new ones to address your specific testing goals.

What diagnostic data adapters are available?

The following list describes the various data and diagnostic adapters that are available for you to configure in your test settings:

- **Actions**: You can create a test setting that collects a text description of each action that is performed during a test. When you configure this adapter, the selections are also used if you create an action recording when you run a manual test. The action logs and action recordings are saved together with the test results for the test. You can play back the action recording later to fast-forward through your test, or you can view the action log to see what actions were taken.

Manual tests (Local machine)	Manual tests (Collecting data using a set of roles and an environment)	Automated tests
Yes	Yes	No

Note: *When you collect data on a remote environment, the recording will work only on the local machine.*

- **ASP.NET Client Proxy for IntelliTrace and Test Impact**: This proxy allows you to collect information about the HTTP calls from a client to a web server for the IntelliTrace and Test Impact diagnostic data adapters.

Manual tests (Local machine)	Manual tests (Collecting data using a set of roles and an environment)	Automated tests
Yes	Yes	Yes

- **ASP.NET profiler**: You can create a test setting that includes ASP.NET profiling, which collects performance data on ASP.NET web applications.

Manual tests (Local machine)	Manual tests (Collecting data using a set of roles and an environment)	Automated tests
No	No	Yes

Note: *This diagnostic data adapter is supported only when you run load tests from Visual Studio.*

- **Code coverage**: You can create a test setting that includes code coverage information that is used to investigate how much of your code is covered by tests.

Manual tests (Local machine)	Manual tests (Collecting data using a set of roles and an environment)	Automated tests
No	No	Yes

Note: *You can use code coverage only when you run an automated test from Visual Studio or mstest.exe, and only from the machine that runs the test. Remote collection is not supported.*

Note: *Collecting code coverage data does not work if you also have the test setting configured to collect IntelliTrace information.*

- **IntelliTrace**: You can configure the diagnostic data adapter for IntelliTrace to collect specific diagnostic trace information to help isolate bugs that are difficult to reproduce. This adapter creates an IntelliTrace file that has an extension of .iTrace that contains this information. When a test fails, you can create a bug. The IntelliTrace file that is saved with the test results is automatically linked to this bug. The data that is collected in the IntelliTrace file increases debugging productivity by reducing the time that is required to reproduce and diagnose an error in the code. From this IntelliTrace file, the local session can be simulated on another computer; this reduces the possibility of a bug being non-reproducible.

Manual tests (Local machine)	Manual tests (Collecting data using a set of roles and an environment)	Automated tests
Yes	Yes	Yes

For more details and the steps used to add and configure the IntelliTrace diagnostic data adapter, see the MSDN topic *How to: Collect IntelliTrace Data to Help Debug Difficult Issues*.

- **Event log**: You can configure a test setting to include event log collecting, which will be included in the test results.

Manual tests (Local machine)	Manual tests (Collecting data using a set of roles and an environment)	Automated tests
Yes	Yes	Yes

To see the procedure used to add and configure the event log diagnostic data adapter, see the MSDN topic *How to: Configure Event Log Collection Using Test Settings*.

- **Network emulation**: You can specify that you want to place an artificial network load on your test using a test setting. Network emulation affects the communication to and from the machine by emulating a particular network connection speed, such as dial-up.

Manual tests (Local machine)	Manual tests (Collecting data using a set of roles and an environment)	Automated tests
Yes	Yes	Yes

For more information about the network emulation diagnostic data adapter, see the MSDN topic *How to: Configure Network Emulation Using Test Settings*.

- **System information**: A test setting can be set up to include the system information about the machine that the test is run on. The system information is specified in the test results by using a test setting.

Manual tests (Local machine)	Manual tests (Collecting data using a set of roles and an environment)	Automated tests
Yes	Yes	Yes

- **Test impact**: You can collect information about which methods of your application's code were used when a test case was running. This information can be used together with changes to the application code made by developers to determine which tests were impacted by those development changes.

Manual tests (Local machine)	Manual tests (Collecting data using a set of roles and an environment)	Automated tests
Yes	Yes	Yes

Note: *If you are collecting test impact data for a web client role, you must also select the ASP.NET Client Proxy for IntelliTrace and Test Impact diagnostic data adapter.*

Note: *Only the following versions of Internet Information Services (IIS) are supported: IIS 6.0, IIS 7.0 and IIS 7.5.*

For further details, see the MSDN topic *How to: Collect Data to Check Which Tests Should be Run After Code Changes*.

- **Video recorder**: You can create a video recording of your desktop session when you run an automated test. This video recording can be useful for viewing the user actions for a coded UI test. The video recording can help other team members isolate application issues that are difficult to reproduce.

Manual tests (Local machine)	Manual tests (Collecting data using a set of roles and an environment)	Automated tests
Yes	Yes	Yes

Note: *If you enable the test agent software to run as a process instead of a service, you can create a video recording when you run automated tests.*

For more information, see the MSDN topic *How to: Record a Video of Your Desktop as You Run Tests Using Test Settings*.

Tip: *The data that some of the diagnostic data adapters capture can take up a lot of database space over time. By default, the administrator of the database used for Visual Studio 2010 cannot control what data gets attached as part of test runs. For example, there are no policy settings that can limit the size of the data captured and there is no retention policy to determine how long to hold this data before initiating a cleanup. To help with this issue, you can download the **Test Attachment Cleaner for Visual Studio Ultimate 2010 & Test Professional 2010**. The test attachment cleaner tool allows you to determine how much database space each set of diagnostic data captures is using and reclaim the space for runs that are no longer relevant to your team.*

Load testing

To help determine how well your application responds to different levels of usage, your team can create load tests. These load tests contain a specified set of your web performance tests, unit tests, or coded UI tests. Load tests can be modeled to test the expected usage of a software program by simulating multiple users who access the program at the same time. Load tests can also be used to test the unexpected!

Load tests can be used in several different types of testing:

- **Smoke**: To test how your application performs under light loads for short durations.
- **Stress**: To determine if the application will run successfully for a sustained duration under heavy load.
- **Performance**: To determine how responsive your application is.
- **Capacity planning**: To test how your application performs at various capacities.

Visual Studio Ultimate lets you simulate an unlimited number of virtual users on a local load test run. In a load test, the load pattern properties specify how the simulated user load is adjusted during a load test. Visual Studio Ultimate provides three built-in load patterns: constant, step, and goal-based. You choose the load pattern and adjust the properties to appropriate levels for your load test goals. For more about load patterns, see the MSDN topic *Editing Load Patterns to Model Virtual User Activities*.

If your application is expected to have heavy usage—for example, thousands of users at the same time—you will need multiple computers to generate enough load. To achieve this, you can set up a group of computers that would consist of one or more test controllers and one or more test agents. A test agent runs tests and can generate simulated load. The test controller coordinates the test agents and collects the test results. For more information about how to set up test controllers and test agents for load testing, see the MSDN topic *Distributing Load Tests Runs across Multiple Test Machines Using Test Controllers and Test Agents*.

Web performance tests in load tests

When you add web performance tests to a load test, you simulate multiple users opening simultaneous connections to a server and making multiple HTTP requests. You can set properties on load tests that will be applied to all of the individual web performance tests.

Unit tests in load tests

Use unit tests in a load test to exercise a server through an API. Typically, this is for servers that are accessed through thick clients or other server services rather than a browser. One example is a Windows application with a Windows Forms or Windows Presentation Foundation (WPF) front end, using Windows Communication Foundation (WCF) to communicate to the server. In this case, you develop unit tests that call WCF. Another example is a different server that calls the server through web services. Additionally, it is possible that a two-tier client makes calls directly to SQL Server. In this case, you can develop unit tests to call SQL Server directly.

Coded UI test in load tests

Load tests can also include automated coded UI tests. The inclusion of coded UI tests should only be done under specific circumstances. All the scenarios that use coded UI tests in load tests involve using the coded UI tests as performance tests. This can be useful because coded UI tests let you capture performance at the UI layer. For example, if you have an application that takes one second to return data to the client but eight seconds to render the data in the browser, you cannot capture this type of performance problem by using a web performance test.

You would also benefit from using coded UI tests in a load test if you have an application that is difficult to script at the protocol layer. In this case, you might consider temporarily driving load using coded UI until you can correctly script the protocol layer.

For more information, see the MSDN topic *Using Coded UI Tests in Load Tests*.

Creating load tests

A load test is created by using the New Load Test Wizard in Visual Studio Ultimate. When you use the New Load Test Wizard, you specify the following three settings for the load test:

- **The initial scenario for the load test**: Load tests contain scenarios, which contain web performance tests, unit tests, and coded UI tests. A scenario is a container within a load test where you specify load pattern, test mix model, test mix, network mix, and web browser mix. Scenarios are important because they give you flexibility in configuring test characteristics that allow for simulation of complex, realistic workloads. You can also specify other various load test scenario properties to meet your specific load testing requirements; for example, delays and think times.

    Tip: *For a list of the load test scenario properties you can modify using the Load Test Editor, see the MSDN topic Load Test Scenario Properties.*

    You can think of a scenario as representing a particular group of users. The tests in the scenario represent the activity of those users, the load pattern is the number of users in the group, and the network and browser settings control the networks and browsers you expect those users to use.

- **Computers and counter sets in the load test**: Counter sets are a set of system performance counters that are useful to monitor during a load test. Counter sets are organized by technology; for example, ASP.NET or SQL counter sets. When you create the load test, you specify which computers and their counter sets to include in the load test.

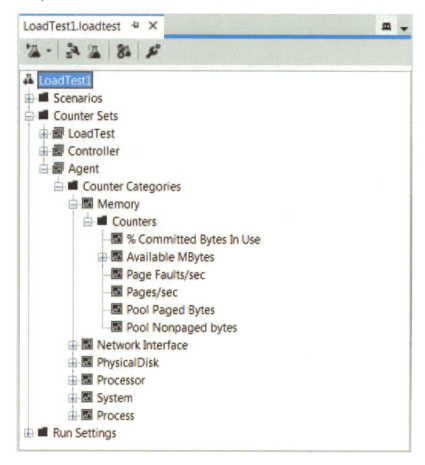

Load test counter sets

> **Note:** *If your load tests are distributed across remote machines, controller and agent counters are mapped to the controller and agent counter sets. For more information about how to use remote machines in your load test, see Distributing Load Test Runs Across Multiple Test Machines Using Test Controllers and Test Agents.*

- **The initial run setting for the load test**: Run settings are a set of properties that influence the way a load test runs.
  You can have more than one run setting in a load test. Only one of the run settings may be active for a load test run. The other run settings provide a quick way to select an alternative setting to use for subsequent test runs.

Tip: *For a list of the run setting properties you can modify using the Load Test Editor, see* **Load Test Run Setting Properties.**

To see the detailed steps that are used in the Load Test Wizard, see the MSDN topic *Creating Load Tests Using the New Load Test Wizard.* The initial settings that you configure for a load test using the New Load Test Wizard can be edited later using the Load Test Editor. For more information, see *Editing Load Test Scenarios Using the Load Test Editor.*

RUNNING AND ANALYZING LOAD TESTS

You view both running load tests and completed load tests in the Load Test Analyzer.

Tip: *Before you run a load test, make sure that all the web performance, unit tests, and coded UI tests that are contained in the load test will pass when they are run by themselves.*

While a test is running, a condensed set of the performance counter data that can be monitored in the Load Test Analyzer is maintained in memory. To prevent the resulting memory requirements from growing unbounded, a maximum of 200 samples for each performance counter is maintained. This includes 100 evenly spaced samples that span the run's current elapsed time, and the most recent 100 samples. The result that is accumulated during a run is called an *in-progress load test* result.

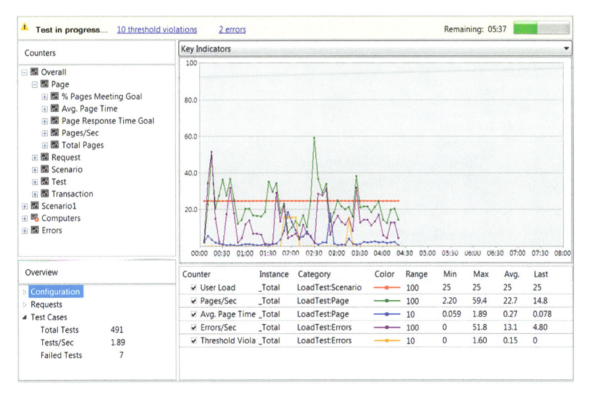

Analyzing a running load test

In addition to the condensed set of performance counter data, the Load Test Analyzer has the following functionality available to analyze the in-progress load test result data that is unique while a load test is running:

- A progress indicator specifies the time that remains.
- A button on the Load Test Analyzer toolbar is available to stop the load test.
- You can specify either collapsing or scrolling graphing modes on the Load Test Analyzer toolbar: Collapsing is the default graph mode in the Load Test Analyzer during a running load test. A collapsing graph is used for load test while it is running to reduce the amount of data that must be maintained in memory, while still showing the trend for a performance counter over the full run duration.
  Scrolling graph mode is available when you are viewing the result of a load test while it is running. A scrolling graph is an optional view which shows the most recent data points. Use a scrolling graph to view only the most recent 100 data intervals in the test.
- An Overview pane which displays the configuration, requests, and test cases information for the running load test.

Running load tests

1. In Visual Studio Ultimate, in your solution, locate your test project and open your load test.
2. In the Load Test Editor, click the **Run** button on the toolbar.

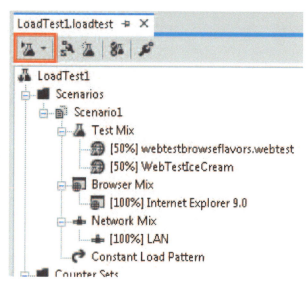

Run load tests

For more information, see the two MSDN topics: *How to: Run a Load Test and Analyzing Load Test Runs.*

IntelliTrace

Undoubtedly, non-reproducible bugs have long been a problem for the developers on your team, as they have been for Contoso. They sometimes saw their applications crash during a test on a test computer, but run without any issues on their developer's computer.

Diagnosing application issues under such circumstances has been very difficult, expensive, and time-consuming for Contoso. The bugs that their developers received likely did not include the steps to reproduce the problem. Even if bugs included the steps, the problem might stem from the specific environment in which the application is being tested.

Fabrikam dealt with this sort of issue by collecting IntelliTrace data in their tests to assist in solving a lot of their non-reproducible errors. Tests configured with a test setting that uses the IntelliTrace diagnostic data adapter can automatically collect IntelliTrace data. The collected data is saved as an IntelliTrace recording that can later be opened by developers using Visual Studio. Team Foundation Server work items provide a convenient means for testers to share IntelliTrace recordings with developers. The developer can debug the problem in a manner similar to postmortem debugging of a dump file, but with more information.

Configuring the IntelliTrace diagnostic data adapter

You can configure the test settings for either Microsoft Test Manager or Visual Studio to use the diagnostic data adapter for IntelliTrace to collect specific diagnostic trace information. Tests can use this adapter, the test can collect significant diagnostic events for the application that a developer can use later to trace through the code to find the cause of a bug. The diagnostic data adapter for IntelliTrace can be used for either manual or automated tests.

> **Note:** *IntelliTrace works only on an application that is written in managed code. If you are testing a web application that uses a browser as a client, you should not enable IntelliTrace for the client in your test settings because no managed code is available to trace. In this case, you may want to set up an environment and collect IntelliTrace data remotely on your web server.*

When you configure the IntelliTrace adapter, you can configure it to collect IntelliTrace events only. When the adapter is configured to collect IntelliTrace events important diagnostic events are captured with minimal impact on the performance of your tests. The types of events that can be collected by IntelliTrace include the following:

- **Debugger events**. These are events that occur within the Visual Studio Debugger while you debug your application. The startup of your application is one debugger event. Other debugger events are *stopping events*, which are events that cause your application to enter a break state. Hitting a breakpoint, hitting a tracepoint, or executing a **Step** command are examples of stopping events. For performance reasons, IntelliTrace does not collect all possible values for every debugger event. Instead, IntelliTrace collects values that are visible to the user. If the Autos window is open, for example, IntelliTrace collects values that are visible in the Autos window. If the Autos window is closed, those values are not collected. If you point to a variable in a source window, the value that appears in the DataTip is collected. Values in a pinned DataTip are not collected, however.

- **Exception events**. These occur for handled exceptions, at the points where the exception is thrown and caught, and for unhandled exceptions. IntelliTrace collects the type of exception and the exception message.
- **Framework events**. These occur within the Microsoft .NET Framework library. You can view a complete list of .NET events that can be collected on the **IntelliTrace Events** page of the **Options** dialog box. The data collected by IntelliTrace varies by event. For a **File Access** event, IntelliTrace collects the name of the file; for a **Check Checkbox**, it collects the checkbox state and text; and so on.

Alternatively, you can configure the IntelliTrace adapter to record both the IntelliTrace events, and method level tracing; however, doing so might impact the performance of your tests. Some additional configuration options for the IntelliTrace diagnostic data adapter include:

- Collection of data from ASP.NET applications that are running on IIS.
- Turning collection of IntelliTrace information on or off for specific modules. This ability is useful because certain modules might not be interesting for debugging purposes. For example, you might be debugging a solution that includes legacy DLL projects that are well tested and thoroughly debugged. Excluding modules that do not interest you reduces clutter in the IntelliTrace window and makes it easier to concentrate on interesting code. It can also improve performance and reduce the disk space that is used by the log file. The difference can be significant if you have chosen to collect calls and parameters.
- The amount of disk space to use for the recording.

To view the detailed steps that you use to add the IntelliTrace Diagnostic Data adapter to your test settings, see the MSDN topic *How to: Collect IntelliTrace Data to Help Debug Difficult Issues*.

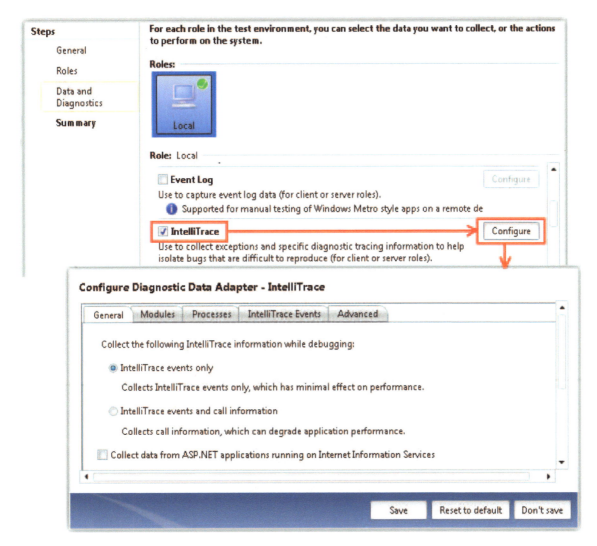

Configure the IntelliTrace Data and Diagnostic Adapter

Fixing non-reproducible bugs with IntelliTrace

An IntelliTrace recording provides a timeline of events that occurred during the execution of an application. Using an IntelliTrace recording, you can view events that occurred early in the application run, in addition to the final state. In this way, debugging an IntelliTrace recording resembles debugging a live application more than it resembles debugging a dump file.

IntelliTrace lets the developers on your team debug errors and crashes that would otherwise be non-reproducible. The developers can debug log files that were created by configuring the IntelliTrace data and diagnostic adapter locally, or from Test Manager. Members of your team can link a log file from Test Manager directly to a Team Foundation Server work item or bug, which can be assigned to a developer. In this way, IntelliTrace and Test Manager integrate into your team workflow.

When you debug an IntelliTrace file, the process is similar to debugging a dump file. However, IntelliTrace files provide much more information than traditional dump files. A dump file provides a snapshot of an application's state at one moment in time, usually just when it crashed. With IntelliTrace, you can rewind the history to see the state of the application and events that occurred earlier in the application run. This makes debugging from a log file faster and easier than debugging from a dump file.

IntelliTrace can increase your cycle time significantly by alleviating time-consuming non-reproducible bugs.

To see the steps used to debug an IntelliTrace recording attached to a bug, see the MSDN topic *Debugging Non-Reproducible Errors With IntelliTrace*.

Feedback tool

Getting the right feedback at the right time from the right individuals can determine the success or failure of a project or application. Frequent and continuous feedback from stakeholders supports teams in building the experiences that will delight customers. As stakeholders work with a solution, they understand the problem better and are able to envision improved ways of solving it. As a member of the team developing the application, you can make course corrections throughout the cycle. These course corrections can come from both negative and positive feedback your team receives from its stakeholders.

The Team Foundation Server tools for managing stakeholder feedback enable teams to engage stakeholders to provide frequent and continuous feedback. The feedback request form provides a flexible interface to specify the focus and items that you want to get feedback about. Use this tool to request feedback about a web or downloadable application that you're working on for a future release. With Microsoft Feedback Client, stakeholders can directly interact with working software while recording rich and usable data for the team in the background through action scripts, annotations, screenshots, and video or audio recordings.

For more information, see the MSDN topic *Engaging Stakeholders through Continuous Feedback*.

Create a feedback request

Creating a feedback request for the stakeholders, customers, or team members involved in the current application cycle is relatively simple. Creating a feedback request can be accomplished using Web Access using the following steps:

1. Connect to Team Web Access by opening a web browser and entering the URL. For example: **http://Fabrikam:8080/tfs/**.
2. On the **Home page**, expand the team project collection and choose your team project.
3. On the **Home page** for the team project, choose the **Request feedback** link.
4. The **REQUEST FEEDBACK** dialog box opens.
5. Follow the instructions provided and fill out the form.

Provide feedback

Your stakeholders respond to your teams request for feedback by using the Microsoft Feedback Client. This tool allows your stakeholders to launch the application under development, capture their interaction with it as video and verbal or type-written comments as well. The feedback is stored in Visual Studio 2012 Team Foundation Server to support traceability. Stakeholders can record their interactions with the application, record verbal comments, enter text, clip and annotate a screen, or attach a file.

When your stakeholders receive an email request for feedback, it contains a link to start the feedback session.

> **Note:** *The email also includes a link to install the feedback tool if it is not already installed.*

We want your feedback for the following items:

1. Editing the ice cream flavors
2. Sorting the ice cream flavors by region
3. Editing the users profile for ice cream favorites

Start your feedback session

If the feedback tool is not already installed on your machine, install the feedback tool.

Thanks,
Adam Barr (Fabrikam)

Email requesting feedback

Clicking the link opens the feedback client on the **Start** page. From the Start page, choose the **Application** link to open, start, or install the application for which you have been requested to provide feedback.

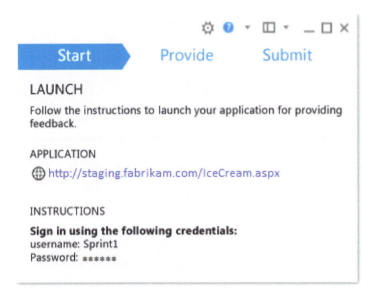

Starting the application

On the **Provide** page, one or more items appear for you to provide feedback. For each item, you can get context on what's being asked and then you can give feedback free-form through video or audio recordings, text, screenshot, or file attachments. When finished with one item, choose the **Next** button to move to the next item.

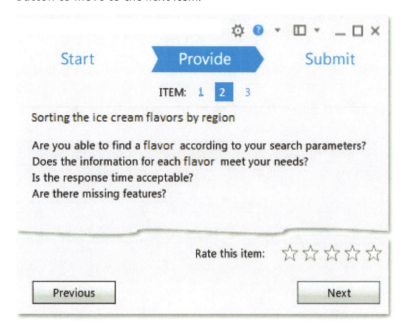

Providing feedback for each item

When providing feedback, your stakeholders can add rich text comments and add screenshots and attach related files. While providing feedback, you can optionally choose to record the feedback session using either **Screen & Voice**, **Screen only**, or **Voice only**.

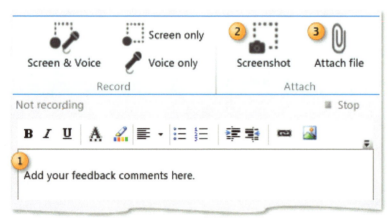

Feedback options

After entering the feedback, stakeholders can easily submit their feedback. The feedback is uploaded to your team project as a work item.

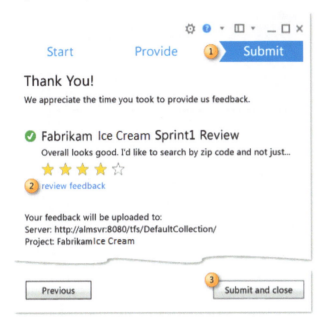

Submit the feedback

Remote debugging

At times, it can be helpful to isolate issues on other machines or devices on your network. For example, you might identify the need to debug an issue on a staging server or test machine that your team is having trouble isolating or replicating. Using the remote debugger, you can conduct debugging remotely on a machine that is in your network. When you are doing remote debugging, the host computer can be any platform that supports Visual Studio.

> **Note:** *The remote device and the Visual Studio computer must be connected over a network or connected directly through an Ethernet cable. Debugging over the Internet is not supported.*

The remote machine or device must have the remote debugging components installed and running. Additionally, you must be an administrator to install the remote debugger on the remote device. Also, in order to communicate with the remote debugger, you must have user access to the remote device. For information on installing the remote debugger, see the MSDN topic *How to: Set Up Remote Debugging*.

Summary

The print statement is no longer the tester's only device. In Visual Studio 2012, there are powerful well-integrated tools. In this chapter we've looked at:

- Performance and stress testing
- Load testing
- IntelliTrace
- Stakeholder feedback tool
- Remote debugging

Differences between Visual Studio 2010 and Visual Studio 2012

- **Creating load and web performance tests**: Load tests and web performance tests are created by adding them to a web performance and load test project in Visual Studio 2012. In Visual Studio 2010, load and web performance tests are created by adding them to a test project. The test project in Visual Studio 2010 was also used for unit tests, coded UI tests, and generic and ordered tests. For more information, see the MSDN topic *Upgrading Web Performance and Load Tests from Visual Studio 2010.*

- **Running load and web performance tests**: In Visual Studio 2012, load tests must be run from the Load Test Editor. Similarly, web performance tests must be run from the Web Performance Test Editor. In Visual Studio 2010, web performance tests and load tests can be run from either their respective editors, or from the Test View window or Test List Editor window.
  In Visual Studio 2012, the **Test** menu that was in Visual Studio 2010 has also been deprecated. To run or debug your coded web performance tests, you must do so from the shortcut menu in the editor. For more information, see the MSDN topic *How to: Run a Coded Web Performance Test.*
  In Visual Studio 2012, the Test View window has been replaced by the Unit Test Explorer, which provides for a more agile testing experience for code development for unit tests and coded UI tests. The Unit Test Explorer does not include support for web performance and load tests.

- **Virtual user limitations for load testing**: Visual Studio Ultimate 2012 includes unlimited virtual users that you can use with your load tests. With Visual Studio Ultimate 2010, you are restricted to 250 virtual users on a local load test run. If your load testing requires more virtual users, or you want to use remote machines, you must install a Visual Studio Load Test Virtual User Pack 2010 or Visual Studio 2010 Load Test Feature Pack.
  You can purchase Visual Studio Load Test Virtual User Pack 2010 where you purchased Visual Studio Ultimate. Each user pack adds an additional 1000 virtual users that are configured on your test controller allowing for running your load tests on virtual machines in your environment.
  The Visual Studio 2010 Load Test Feature Pack is available if you are an MSDN subscriber. The feature pack provides unlimited virtual users! Another benefit from installing either the unlimited virtual user license in this feature pack or Visual Studio Load Test Virtual User Pack 2010 is that they enable multiprocessor architecture. The multiprocessor architecture allows the machine that the licenses are installed on to use more than one processor. Otherwise, the machine is restricted to using only one core.

- **Upgrading test controllers used with load tests or web performance tests**: If you are using test controllers from Visual Studio for web performance or load testing—these test controllers are not configured with Team Foundation Server—then the version of test controller must match the version of Visual Studio. For more information, see the MSDN topics *Upgrading Test Controllers from Visual Studio 2010* and *Installing and Configuring Test Agents and Test Controllers*.

- **Feedback client**: The Feedback Client tool is new for Visual Studio 2012 and did not exist for Visual Studio 2010.

- **Remote debugger**: The Visual Studio remote debugging process has been simplified. Installing and running the remote debugger no longer requires manual firewall configuration on either the remote computer or the computer running Visual Studio. You can easily discover and connect to computers that are running the remote debugger by using the **Select Remote Debugger Connection** dialog box.

Where to go for more information

All links in this book are accessible from the book's online bibliography available on MSDN: *http://msdn.microsoft.com/en-us/library/jj159339.aspx*.

7 Testing in the Software Lifecycle

Testing is a vital part of software development, and it is important to start it as early as possible, and to make testing a part of the process of deciding requirements. To get the most useful perspective on your development project, it is worthwhile devoting some thought to the entire lifecycle including how feedback from users will influence the future of the application. The tools and techniques we've discussed in this book should help your team to be more responsive to changes without extra cost, despite the necessarily wide variety of different development processes. Nevertheless, new tools and process improvements should be adopted gradually, assessing the results after each step.

Testing is part of a lifecycle. The software development lifecycle is one in which you hear of a need, you write some code to fulfill it, and then you check to see whether you have pleased the stakeholders—the users, owners, and other people who have an interest in what the software does. Hopefully they like it, but would also like some additions or changes, so you update or augment your code; and so the cycle continues. This cycle might happen every few days, as it does in Fabrikam's ice cream vending project, or every few years, as it does in Contoso's carefully specified and tested healthcare support system.

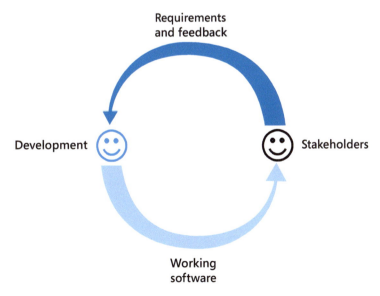

Software development lifecycle

Testing is a proxy for the customer. You could conceivably do your testing by releasing it into the wild and waiting for the complaints and compliments to come back. Some companies have been accused of having such a strategy as their business model even before it became fashionable. But on the whole, the books are better balanced by trying to make sure that the software will satisfy the customer before we hand it over.

We therefore design tests based on the stakeholders' needs, and run the tests before the product reaches the users. Preferably well before then, so as not to waste our time working on something that isn't going to do the job.

In this light, two important principles become clear:

- **Tests represent requirements**. Whether you write user stories on sticky notes on the wall, or use cases in a big thick document, your tests should be derived from and linked to those requirements. And as we've said, devising tests is a good vehicle for discussing the requirements.
- **We're not done till the tests pass**. The only useful measure of completion is when tests have been performed successfully.

Those principles apply no matter how you develop your software.

Process wars

Different teams have different processes, and often for good reasons. The software that controls a car's engine is critical to its users' safety and difficult to change in the field. By contrast, a vending site is less obviously safety-critical, and can be changed in hours. The critical software is developed with high ceremony—much auditing, many documents, many stages and roles—while the rapid-cycling software is developed by much less formal teams with less differentiated roles and an understandable abhorrence of documentation, specifications, and book chapters about process.

Nevertheless, development teams at different locations on this axis do have some common ground and can benefit from many of the same tools and practices.

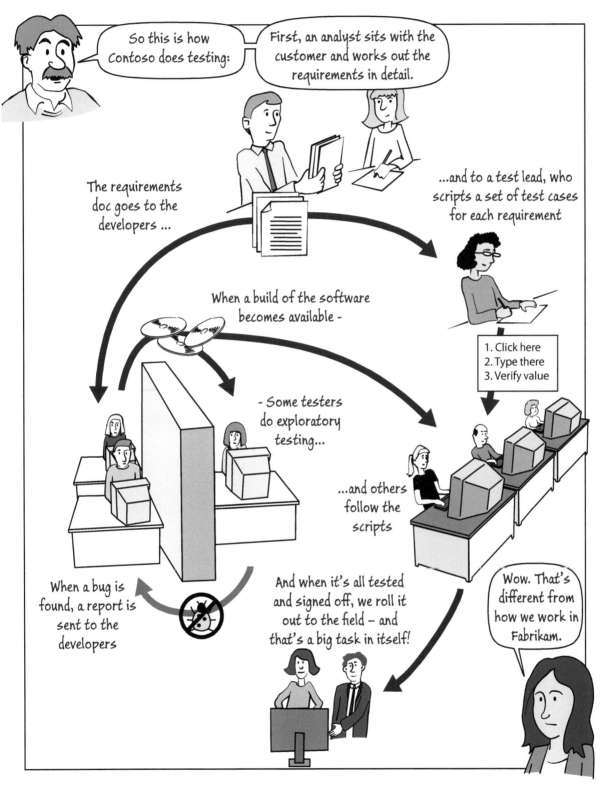

Our cycle from client's wishes to deployed code is much shorter. We don't have all those separate stages.

And we don't have separate roles. Sure, we have some really good bug finders, and some really great coders, and specialist graphic designers; but everybody does some of each.

We often demo to the client, and discuss improvements. For cloud apps we release each feature as it's developed, and see what feedback we get. We don't do big specification upfront, and expect new ideas to come up as we go along. What we end up with is just exactly what the users need.

No <u>way</u> would that ever be the way we work! Not for our healthcare business, anyway.

We are <u>not</u> going to determine requirements by trying out wacky ideas on the live hospital system every couple of weeks!

But your process is so heavyweight...

WHO CREATES SYSTEM TESTS?

There are two sides to the debate over whether system tests should be performed by the developers, or by a separate group of testers. As usual, the answer is somewhere in between the extremes, and depends on your circumstances.

In projects where a rapid overall cycle is essential, a separation between development and test is not favored. It introduces a handoff delay. Bugs aren't detected until the developers have moved on.

In support of separation, the most compelling argument is that testing and development require different skills and different attitudes. While developing, you think "How can I make this as useful as it can be for its users?" but in testing you think "I am going to break this!" Members of a development team—so the argument runs—tend to be too gentle with the product, moving it nicely along the happy paths; while the skilled tester will look for the vulnerabilities, setting up states and performing sequences of actions which the developers would find perverse in order to expose a crack in the armor.

The most versatile testers are those who can write code; just as the most versatile developers are those who can devise a good test. In Microsoft we have a job title, Software Development Engineer in Test. They are specialists whose skills go beyond simply trying out the software, and also beyond simply writing good code. A good SDET is able to create model-based tests; and has sufficient insight into the design of the system and its dependencies to be able to spot the likely sources of error.

Whether you have separate testers depends on the history of your team, and the nature of your project. We don't recommend separate development and test teams, unless you're writing a critical system. Even if different individuals focus on development and test, keep them on the same team. Move people occasionally between the two roles.

If you do have separate testers, bear in mind the following points:

- Writing test cases is a crucial part of determining requirements. Make sure your test team members are key participants in every discussion of user stories (or other requirements). As we discussed in Chapter 4, "Manual System Tests," some teams practice acceptance test-driven development, in which the system tests are the only statement of the requirements.
- Automated system tests are important tools for reducing regression bugs, and an important enabler for shortening your development cycles. Introduce a developer or two to the test team. Our experience is that this brings substantial improvements in many aspects of testing.

DevOps

An application spends most of the lifecycle deployed for customers to use it. Although the excitement and challenge in developing a software system is intense, from an operational perspective it's just part of the early lifecycle of the system. What the operational folks care more about is how well behaved the system is in actual operation.

One thing you can be certain of: No matter what you do, there will nearly always be another version, another release, another iteration. Your client, flush with their initial success, will stretch their ambition to selling crêpes as well as ice cream. Or they will discover they can't add flavors that contain accented characters (and consider this to be a bug, even though they never asked). And, yes, there will be bugs that find their way through to the working system.

By contrast, in a typical web sales application, the cycle should be very short, and updates are typically frequent. This is the sort of process Fabrikam had finely honed over the last few years. The time that elapses between a new business idea such as "let's let customers set up a weekly order" and implementing it in the software should be short because it's not only software developers who practice agility. A successful business performs continuous small improvements, trying out an idea, getting customer feedback, and making further changes in response.

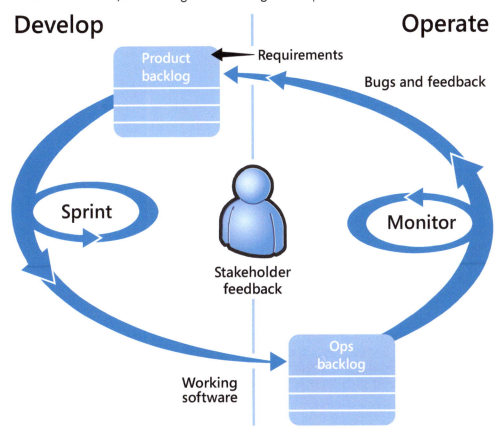

Continuous improvement

In the aircraft project, there are many iterations of development between each release of working software; in the web project, there are fewer—or maybe none. In *continuous deployment*, each feature released as soon as it is done.

The study of this cycle has come to be called "devOps." Like so many methodological insights, the point is not that we've just invented the devOps cycle; software systems have always had next versions and updates, whether the cycles take years or hours. But by making the observation, we can evaluate our objectives in terms of a perspective that doesn't terminate at the point of the software's delivery.

For example, we can think about the extent to which we are testing not just the software, but the business process that surrounds it; and what we can do to monitor the software while it is operational; and how we adopt software that is already in a devOps cycle.

Testing the business process

Don't forget that your application is just a part of something that your users are doing, and that their real requirements are about that process. Whether they are making friends, buying and delivering ice cream, or running an oil refinery, the most important tests are not about what's displayed on the screen, but about whether the friends are kept or offended, the ice cream is delivered or melted, and the oil cracked or sent up in flames.

Testing critical or embedded software involves setting up test harnesses that simulate the operation of the factory, aircraft, or phone (as examples). The harness takes the system through many sequences of operation and verifies that it remains within its correct operational envelope. This is outside the scope of this book, beyond noting that simulation isolates the system just as we discussed isolating units in Chapter 2, "Unit Testing: Testing the Inside."

For less critical systems, and tools that depend heavily on the behavior and preferences of their users, the only real way to test the process surrounding your application is to observe it in operation.

Operational monitoring

There are two aspects to operational monitoring: monitoring your system to see if it behaves as it should; and monitoring the users to see what they do with it. What you want to monitor depends on the system. A benefit of bringing the testing approach into the requirements discussions is a greater awareness of the need to design monitoring into the system from the start.

If your system is a website on Internet Information Services (IIS), then you can use the IntelliTrace feature of Visual Studio to log a trace of method calls and other events while it is running. For the default settings, the effect on performance is small, and you can use the trace to debug any problems that are seen in operation. To use this, download the standalone *IntelliTrace Collector*. You can use Visual Studio Ultimate to analyze the trace later.

Shortening the devOps cycle

Testing allows you to go around the cycle more rapidly. Instead of consulting the customers, you can push a button and run the tests.

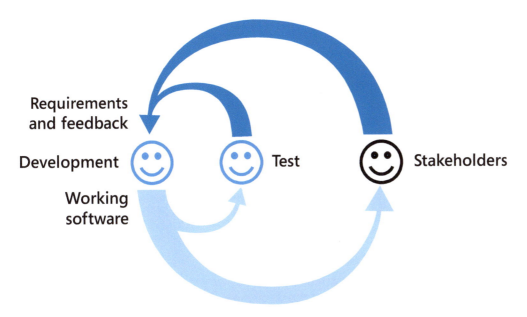

Rapid devOps cycle with test as a proxy for stakeholders

Of course, tests are no real substitute for letting the clients try the software. But tests can do two things more effectively. Automated tests (and, to a lesser extent, scripted tests) can check very rapidly that nothing that used to work has stopped working. Manual tests—especially if performed by an experienced tester—can discover bugs that ordinary clients would take much longer to find, or which would be expensive if they appeared in live operation.

Feedback from tests and stakeholders reduces the risk that you are producing the wrong thing. To reduce the waste of going down wrong tracks, run tests as early as possible, and consult with your clients as frequently as possible.

UPDATING EXISTING CODE

If you're updating a system that exists already, we hope there are tests for it and that they all pass. If not, you need to write some. Write tests around the boundary of the part that you want to change. That is, write tests for the behavior that will not change, of the components that will change.

For example, if you're thinking of just replacing a single method with something that does the same job more efficiently, then write a unit test for the behavior that should be exhibited by both old and new methods. If, on the other hand, you're thinking of rewriting the whole system, write system tests for the system features that will not change.

The same approach applies whether you write coded tests or manual test steps.

Make sure the tests pass before you make the changes to the code. Then add tests for the new or improved behavior. After the updates to the code, both sets of tests should pass.

Testing in the development cycle

Methodologies vary, and your own project will be different, as we mentioned. But in any case, there is a role for testing in every phase of your project and throughout the lifetime of the application. Let's now take a look at those roles in at different stages in a project's life.

TESTS ARE EXECUTABLE REQUIREMENTS

No matter what methodology you follow, it's a fundamental truth that the system tests are the requirements—user stories, product backlog items, use cases, whatever you call them, rendered into executable form.

The requirements might change from day to day. Quite likely, they will change when you ask about some fine detail of a test you are writing.

Therefore:

- **Create a test suite from each requirement**. Create test cases to cover a representative sample of cases. In each test case, write the manual test steps or the automated test code to exercise one particular scenario. Parameterize it to allow a variety of inputs.
- **Cover both functional and quality of service (QoS) requirements**. QoS requirements include security, performance, and robustness.
- **Include unstated requirements**. Many of the QoS requirements are not explicitly discussed. Your job as a tester is to make sure these aspects are covered.
- **Discuss the test cases with the stakeholders**. The relationship between requirements and tests is not one-directional. Your need to make precise tests helps to clarify the requirements, and feeds back into them.
- **Explore. Much of the system's behavior was not explicitly specified**. Did the client actually state that they didn't want a picture of a whale displayed when the users buy an ice cream? No. Is it desirable behavior? Perhaps not. Perform exploratory testing to discover what the system does.
- **Automate important tests gradually**. You need to repeat tests for later iterations, later releases, and after bug fixes or updates. By automating tests, you can perform them quickly as part of the daily build. In addition, they're repeatable—that is, they reliably produce the same results each time, unless something changes.

- **Don't delay testing**. Test as soon as the feature is available. Plan to write features in small increments (of no more than a few days) so that the lag between development and testing is short.
- **System tests are the arbiter of "done."** No one goes home until there are automatic or manual tests that cover all the requirements, and all the tests pass. This applies to each feature, to each cycle in your process, and to the whole project.
  ("No one goes home" is a metaphor. Do not actually lock your staff in; to do so may be contrary to fire regulations and other laws in your locality.)

INCEPTION

At or near the inception of the project, sometimes called Sprint 0, you will typically need to:

- **Set up the test infrastructure**. Build the machinery, the services, the service accounts, the permissions, and other headaches, source control, builds, labs, VM templates. See the Appendix, or, (even better) talk to someone who has done it before.
- **Make sure you know what you're doing**. Get the team together and agree on the practices you're going to follow. Start with what everyone knows, and add one practice at a time. If all else fails, make them read this book, and Guckenheimer & Loje as well.
- **Create or import and run tests for existing code**. If you are updating existing code, make sure you have the existing manual or automatic tests. If it was created via Visual Studio application lifecycle management processes, it will be easy.
- **Understand what the stakeholders want**. Spend a lot of time with the clients and developers. Together with them, write down business goals and values; create slide shows of user scenarios; write down user stories; and draw business activity diagrams, models of business entities and relationships, and diagrams of interactions between the principal actors.
- **Understand the architecture**. Spend time with the developers. Work out the principal components of the system, and the dependencies on external systems such as credit card authorities. Discuss how each component will be tested in isolation as it is developed. Understanding how the system is built tells you about its vulnerabilities. As a tester, you are looking for the loose connections and the soft bits.
- **Draft the plan**. The *product backlog* is the list of user stories, in the order in which they will be implemented. Each item is a requirement for which you will create a test suite. At the start of the project, the backlog items are broad, and the ordering is approximate. They are refined as the project progresses.
  Product backlog items (PBIs) are named and described in terms that are meaningful to stakeholders of the project, such as users and owners. They are not described in implementation terms. "As a customer I can order an ice-cream" is good; "A customer can create a record in the order database" is bad.
  Each item also has a rough estimate of its cost in terms of time and other resources. Remind team members that this should include the time taken to write and run unit tests and system tests. (And while you're there, mention that any guidance documents that might be required should be factored in to the cost as well, and that good technical writers don't come cheap. On the other hand, if user interfaces were always as good as they could be, help texts and their writers would arguably be redundant.)

EACH SPRINT

A development plan is typically organized into iterations, mostly called sprints, even in teams where the Scrum methodology has not been adopted wholesale. Each sprint typically lasts a few weeks. At or near the beginning of each sprint, a set of requirements is picked from the top of the product backlog. Each item is discussed and clarified, and a collection of development tasks is created in Team Foundation Server. You create test suites for each requirement.

Early in the sprint, when the developers are all heads-down writing code, testers can:

- Write test cases for system tests. Create test steps. If there are storyboard slide shows, write steps that work through these stories. Writing down the steps in advance is important: it avoids a bias towards what the system actually does.
- Automate some of the manual tests from previous sprints. Automate and extend the important tests that you'll want to repeat. These tests are used to make sure features that worked before haven't been broken by recent changes to the code. Add these tests to the daily builds.

When the developers check in working features:

- Perform exploratory tests of the new features. Exploratory testing is vital to get a feel for how the product works, and to find unexpected behaviors. From exploratory tests, you will usually decide to write some new test cases.
- Perform the manual scripted test cases that you planned for the new requirements.
- Log bugs. (Anyone can log bugs—testers, developers, product owners, and even technical writers.)
- A bug should initially be considered part of what has to be done for the current iteration. If a bug turns out to need substantial work, discuss it with the team and create a product backlog item to fix it. Then you can put that into the usual process of assigning PBIs to iterations.

Towards the end of the sprint:

- Run important manual tests from previous sprints that have not yet been automated, to make sure those features are still working.
- Start automating the most important manual tests from this sprint.

At the end of the sprint:

- System testing is the primary arbiter of "done" for the sprint. The sprint isn't complete until the test cases assigned to that sprint all pass.

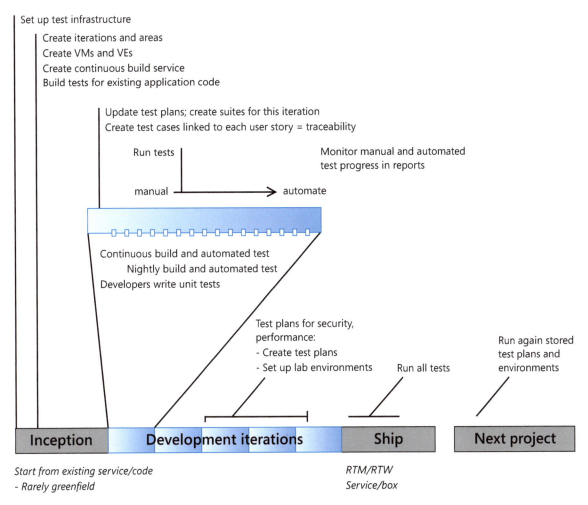

Set up test infrastructure

Create iterations and areas
Create VMs and VEs
Create continuous build service
Build tests for existing application code

Update test plans; create suites for this iteration
Create test cases linked to each user story = traceability

Run tests

Monitor manual and automated
test progress in reports

manual ──────→ automate

Continuous build and automated test
Nightly build and automated test
Developers write unit tests

Test plans for security,
performance:
- Create test plans
- Set up lab environments   Run all tests

Run again stored
test plans and
environments

| Inception | Development iterations | Ship | | Next project |

Start from existing service/code
- Rarely greenfield

RTM/RTW
Service/box

Test activities within sprints

USING PLANS AND SUITES

A test plan represents a combination of test suites, test environment, and test configuration.

Create a new test plan for each iteration of your project. You can copy test suites from one to another.

If you want to run the same tests on different configurations, such as different web browsers, create a test plan for each configuration. Two or more plans can share tests, and again you can copy suites from one plan to another, changing only the configuration of the test environment in the Test Plan's properties.

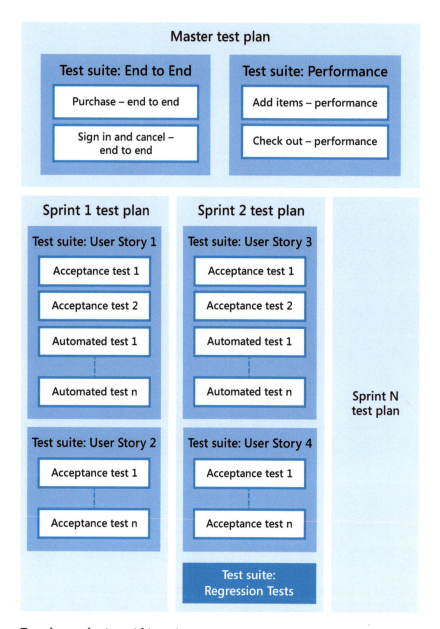

Test plans and suites within sprints

You can, in essence, branch your test plan by performing a clone operation on your test suites. The cloning operation allows your team to work on two different releases simultaneously.

REPORTS

The project website provides a number of graphical reports that are relevant to testing. The choice of reports available, and where you find them, depends on the Team Foundation Server project template that your team uses. See *Process Guidance and Process Templates* and its children on MSDN.

Test plan progress report

This graph summarizes the results of tests in a chosen test plan, over time. The graph relates to one iteration.

Worry if the green section is not increasing. It should be all green towards the end of the iteration.

The total number of test cases is represented by the total height of the graph. If your practice is to write a lot of test cases in advance, you should see a sharp rise followed by a relatively flat section.

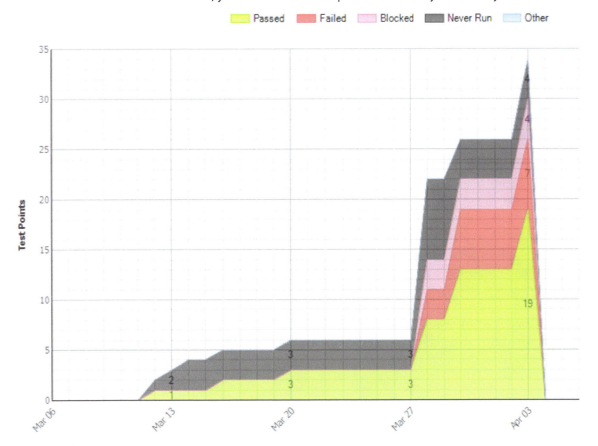

Number of test points

User story test status

The user story test status report is a list of requirements that are scheduled to be implemented in the current iteration, showing the results of the tests linked to each requirement.

Worry if one of the requirements shows an unusually short line. The requirements should be roughly balanced in terms of how many tests they have—if not, break up the larger ones.

Worry if there is a large red section. By the time the requirements are manually tested, they should be mostly passing.

Worry if the chart isn't mostly green towards the end of the iteration.

Title	Work Progress		Hours Remaining	Test Points	Test Status		Bugs
	% Hours Completed				Test Results		
Customers can buy ice cream.	60 %		16	3	67 %	33 %	3
Customers can select flavor from catalog.	60 %		10	3	67 %	33 %	
Vendor can vary flavor catalog.	25 %		15	2	100 %		
Vendor can set different prices for flavors.	39 %		23	2	100 %		4
Customers can set favorite flavors.	100 %		0	1	100 %		4
Users can choose UK spelling of favourite flavour.	100 %		0	6	50 %	33 %	
Customers can select different types of cones.	75 %		5	22	63 %	26 %	2

User story test status

Test case readiness

When you plan a test case, you can set a flag that says whether it is Planned or Ready. The purpose of this is simply to make it easy to see what stage the tests are in, if the practice on your team is to work out test case steps in advance.

At the start of an iteration, you create test cases from requirements, just with their titles. Then you spend some time working out the steps of the test cases. When the steps are worked out and the code is ready, the tests can be run. This chart shows the state of the test cases for the iteration.

On the other hand, if your practice is to generate most test cases from exploratory tests after the code is ready, you might find this chart less useful.

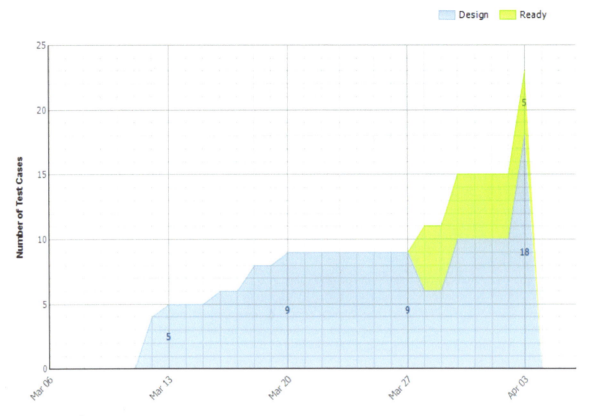

Test case readiness

Process improvements

To summarize the points we've discussed, here are some of the improvements Contoso could consider making to their development process. Any change should be made one step at a time, and assessed carefully before and after.

- When you are discussing scenarios, user stories, or other requirements, think of them sometimes in terms of how you will test them. Tests are executable requirements. Discuss test cases with stakeholders as one way of clarifying the requirements.
- Think about validating not only the system but the business process around it. How will you verify that the user really did get and enjoy the ice cream? How will you use the feedback you get?
- Create requirement and test case work items in your team project, and write out the steps of some key test cases. This provides a clear objective for the development work.
- Write unit tests and aim for code coverage of 70-80%. Try to write at least some of the tests before the application code. Use the tests as a means to think about and discuss what the code should do.

- Use fakes to isolate units so that you can test them even if units they depend on aren't yet complete.
- Set up continuous build-test runs, and switch on email alerts for failing tests. Fix build breaks as soon as they occur. Consider using gated check-ins, which keep the main source free of breaks.
- Plan application development as an iterative process, so that you get basic demonstrable end-to-end functionality early on, and add demonstrable functionality as time goes on. This substantially reduces the risks of developing something that isn't what the users need, and of getting the architecture wrong. Unit tests make it safe to revisit code.
- Set up virtual labs to perform system tests.
- Run each test case as soon as you can. Whenever a build becomes available, perform exploratory testing to find bugs, and to define additional test cases.
- Record your actions when you run manual test cases. In future runs, replay the steps. This allows you to run regression tests much more quickly.
- Use the one-click bug report feature of MTM with environment snapshots, to make bugs easier to reproduce.
- Automate key manual tests and incorporate them into fully automated builds. This provides much more confidence in the system's integrity as you develop it.
- If your team doesn't distinguish the roles of test and development, consider identifying the people who are good at finding bugs, and get them to focus for some of the time on exploratory testing. Ask them also to think about test planning.
- In teams that do separate the developers and testers, consider mixing them up a bit. Involve test leads in the requirements process, and bring developers in to help automate tests.

Thinking of requirements in terms of tests helps to make the requirements more exact

What do we get by testing this way?

Faster response to requirements changes and bug reports.

- Because you have automated many regression tests and can quickly replay others, testing is no longer an obstacle to releasing an update. Manual testing is only required for the new or fixed feature. With virtual lab environments, you can begin testing very quickly. So let's fix that back-button bug, run the regression tests overnight, and release in the morning.

- Stakeholders can experiment and tune their systems. For example, the ice cream vendors need not speculate about whether customers would like to advertise their favorite flavors on a social networking site; they can try the idea and see whether it works.

Reduced costs through less waste.

- Unit tests and automated system tests make it acceptable to revisit existing code because you can be confident that you will find any accidental disturbances of the functions you already have working. This means that instead of finishing each piece of code one at a time, you can develop a very basic working version of your system at an early stage. If your architecture doesn't work as well as you'd hoped, or if the users don't like it as well as they'd thought, you avoid spending time working on the wrong thing. Either you can adjust the project's direction, or, at worst cancel at an early stage.

- Fewer "no repro" bugs. The action recording feature of Microsoft Test Manager automatically logs the steps you took in your bug report. Fewer arguments among team members.

Happier customers.

- By getting regular feedback, you allow your users to try out the system and tune both their process and what they want of the system. The end product is more likely to fit their needs.

- When your stakeholders ask for something different, or when they report a bug, you can improve or fix your system quickly.

What to do next

We hope you've got some ideas from this book that will help you and your team develop software that satisfies your clients better and faster and with less pain to you. Maybe you're like Fabrikam and already have these practices and tools in daily use; maybe you feel your company is more like Contoso and you feel there are a number of improvements that can be made. Most likely, you're like us: as we look around our company, we can see some parts that have very sophisticated testing and release practices, and other parts that are still thinking about it.

Organizations learn and change more slowly than individuals. Partly this is because they need time to achieve consensus. Each person's vision of the way forward might be good and effective if implemented across the team, but chaos comes from everybody following their own path. Partly it's because, with substantial investments at stake, it's important to take one step at a time and assess the results of each step.

You must not only have a vision of where you want to get to, but also form an incremental plan for getting there, in which each step provides value in its own right.

In this book we've tried to address that need. The plan of the book suggests a possible route through the process of taking up the Visual Studio testing tools, and the adaptations in your process that you can achieve along the way.

In this book, we've advocated using the Visual Studio tools to automate your testing process more, and we've shown different ways you can do that. That should make it possible to do regression tests more effectively and repeatably, while freeing up people to do more exploratory testing. We've also recommended you bring testing forward in the process as far as you can to reduce risk, and to make defining tests part of determining what the stakeholders need. In consequence, you should be able to adopt a process that is more responsive to change and satisfies your clients better.

But whatever you do, take one step at a time! The risks inherent in change, and the fear of change, are mitigated by assessing each change as you go.

Bug Bash by Hans Bjordahl

http://www.bugbash.net/

The above cartoon is reproduced with permission from the creator, Hans Bjordahl. See *http://www.bugbash.net/* for more Bug Bash cartoons.

Where to go for more information

All links in this book are accessible from the book's online bibliography available on MSDN: *http://msdn.microsoft.com/en-us/library/jj159339.aspx.*

APPENDIX

Setting up the Infrastructure

In this appendix, we've brought together all the stuff about setting up your environment for application lifecycle management with Visual Studio (with emphasis on the test tools). You won't need all of it when you first start, but we figure that, if you're the team's administrator, you'd prefer to have the instructions all in one place rather than spread throughout the book.

Here's what we'll cover in this appendix:

- **The Testing Infrastructure** – All the software components you'll have to assemble, and how to choose between the different ways of distributing them between boxes.
- **Permissions** – Licensing and service accounts.
- **Installing the Software** – How to install the features of Visual Studio that you will use for application lifecycle management.
- **Populating the VM Library** – How to create new VM templates.

We have provided a couple of popular paths through the forest of options. We suggest you begin with these and then tune things to your own project's needs when you have more familiarity.

You'll see lots of tools and technologies referenced as you read through this appendix. Don't worry; you can always access them by going to the online version of the bibliography. The URL can be found at the end of each chapter.

For a set of hands-on labs and detailed map of even more alternatives, you can download the *Visual Studio Lab Management Guide*. Also see *Installing Team Foundation Components*.

The testing infrastructure

WHAT SOFTWARE DO I NEED?

The features we've been discussing are provided by a combination of components, which together form the testing infrastructure. Here are the products you will use, in approximate priority order. We'll discuss the options in more detail shortly.

We have written this book with Visual Studio 2012 in mind. However, you can do nearly everything we describe with Visual Studio 2010—though sometimes less effectively or conveniently. We will note where there's a significant difference.

We used Windows 7 and Windows Server 2008 to create our examples, but Windows 8 will work just as well. Visual Studio 2012 has features specifically aimed at testing Windows 8 apps, but we don't talk about them in this book.

Product or Feature	Software and Version	For
Visual Studio 2012 or 2010	Visual Studio Professional	Unit testing, debugging. Team Foundation Server UI.
	Visual Studio Premium	All the above, plus code coverage, coded UI tests.
	Visual Studio *Ultimate	All the above, plus web performance, load tests, MTM.
	Visual Studio Test Professional	MTM. Team Foundation Server UI. No unit testing or debugging.
Microsoft Test Manager (MTM)	Installed as part of Visual Studio Ultimate or Test Professional	User interfaces to lab management, test planning, and test runner.
Windows	Windows Server 2008 R2 Enterprise R2 64 bit (includes Microsoft Hyper-V technology)	Servers running the testing infrastructure components. You need a license for each machine, even if it is virtual. (You do not need a license for short-lived test installations.)
	Windows 7	Desktop or laptop machines.
Team Foundation Server	Visual Studio 2012 or 2010 Team Foundation Server	Work item tracking, source control, build service.
System Center Virtual Machine Manager (SCVMM)	Virtual Machine Manager 2008 R2 SP1 or 2012	Virtual machines for test environments.
Microsoft SharePoint Server	*SharePoint Standard; or	Project website with basic reports.
	*SharePoint Enterprise	Extended reports and dashboards. See *Dashboards* on MSDN.
Microsoft SQL Server	SQL Server Express	Supports SCVMM.
	**SQL Server 2008 Standard	Supports Team Foundation Server. Provides reporting services – charts of team progress and product quality.
Key Management Service (KMS)		Provides activation keys to test virtual machines.

* If you need to economize, you can manage without SharePoint.

** You could use SQL Server Express to support Team Foundation Server. However, you should be careful with this decision; you can't easily migrate Team Foundation Server from one SQL Server edition to another, and you will lose the historical data in your work items.

CONFIGURATION CHOICES: SPREADING THE SOFTWARE ON THE HARDWARE

What hardware do you need?

The quick answer is: Two or more big servers; the more the better; but the fewer, the more economical.

One big server is the test host computer. On it, you run virtual machines in which you perform your tests. If you want to run a lot of tests at the same time, you can add more test host servers.

The other server or servers are for the other components of the testing infrastructure: that is, the products such as Team Foundation Server that we've listed in the table above, together with their various auxiliary agents, services, and consoles.

If your project is large, you'll want to spread the infrastructure components out over several computers. To understand that better, let's discuss what the various parts do.

The components of the infrastructure

Whether you pack the components all into one computer, or spread them over many, their functions and relationships are the same:

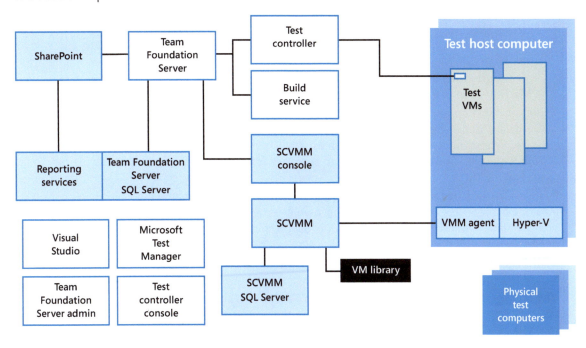

☐ = Must be set up before Team Foundation Server (TFS)

The test infrastructure

In this diagram, all the boxes are software products, except for the test host computer and the physical test computers.

- **Test host** is a physical computer that runs *Hyper-V*. Virtualization allows several virtual machines to share the resources of the physical computer. You can have more than one test host computer so that many virtual machines can run at the same time. Hyper-V is a role of Windows Server that you can switch on in the Computer Management control panel.

  - Test host computers should have a lot of disc space, RAM, and logical CPUs. Virtualization divides these resources between the running virtual machines.

  - We recommend that you don't install any significant service on a test host except Hyper-V. Put the other products on a separate computer or computers.

  - It is possible to run tests without Hyper-V, but we recommend you use it because it makes it much easier to create fresh testing environments.

- **Test virtual machines (VMs)** are machines that your team will create for running tests. Each machine behaves almost exactly like a physical computer. For example, you can register it on your organization's network, start it and shut it down, open the console screen, or log into it with Remote Desktop. In addition, you can save its state and reset it to that state later; you can make duplicate copies of it, and store them in a library.
    - We recommend running tests on VMs because it is very easy to create new VMs with different configurations of virtual hardware, operating systems, and other software. It is also very easy to reset them to a clean state.
    - Most tests can be run on VMs; the exceptions are tests involving special hardware, and some kinds of performance tests.
    - You can use other virtualization frameworks, such as VMware or Citrix XenServer, to run VMs. However, you have to use their facilities to save and restore virtual test machines.
    - Physical test computers are useful for performance tests, or tests that must run on specific hardware. But if this is the first time you've used virtual machines, you'll be surprised how infrequently you need the physical boxes. You'll come to find them inconvenient by comparison with virtual test machines. You can even run some types of performance testing on virtual machines.
- **System Center Virtual Machine Manager (SCVMM)** lets you combine multiple Hyper-V hosts into one big resource. You can create your own private cloud. It also lets you create a library of VM templates from which to create new virtual machines. After installing SCVMM, you don't need to work with the individual Hyper-V consoles.
- **VMM agent** is the proxy for SCVMM that is installed on each test host. Its job is to allow SCVMM to control Hyper-V.
- **VM Library** is a file store for copies and templates of virtual machines. Each virtual machine is typically several gigabytes in size. This file store therefore has to be huge, and should be on a disk by itself. The first VM library must be on the same computer as SCVMM, but you can add further libraries, which can be on separate computers.
- **SCVMM SQL Server** stores information about the stored and running virtual machines.
- **SCVMM console** is the administrator's user interface to SCVMM. You should install a console on the same machine as Team Foundation Server. You can also install consoles on other machines from which you want to control SCVMM. Additionally, you can install a Self-Service Portal for SCVMM, not shown here, which provides a web interface that allows non-admin users to create VMs.
- **Team Foundation Server** stores work items such as test cases, user stories, bugs, and developer tasks.
    - Team Foundation Server also provides version control for source code and other project artifacts.
    - For a big project, you can install instances of Team Foundation Server on separate machines and combine them into one logical cluster, which has a single URI for user access.

- **Lab Manager** allows you to create and manage lab environments, which are groups of virtual or physical machines on which you can run tests. Lab Manager is part of Team Foundation Server, and it provides features on top of SCVMM.
- *Build Service* performs continuous or regular builds on the server, taking its source code from Team Foundation Server source control. It communicates with a build agent in each test machine, allowing it to deploy compiled code into the machine.
- **Test controller** runs the script that performs a test. In each test machine, it works with a test agent to start the tests and collect data.

Client components – typically located on your desktop:

- **Visual Studio** is the principal interface to Team Foundation Server work items, source control, and build service.
- **Microsoft Test Manager (MTM)** is the interface to Lab Manager and the test controller.
- **TFS Admin**, the administrator's console of Team Foundation Server.

Lab Manager and virtualization

We deal with Lab Manager in more depth in Chapter 3, "Lab Environments." Let's take a moment to understand the relationship between Lab Manager and virtualization.

Lab Manager is a feature of Team Foundation Server. You enable it from the Team Foundation Server console. Users access Lab Manager from their desktop machines by using the Lab Center section of Microsoft Test Manager.

Lab Manager lets users create and control lab environments. A lab environment is a group of machines on which you can install a system that is to be tested. In particular, lab environments are designed for testing distributed systems such as web services.

A lab environment can be composed from physical or virtual machines. The big advantages of using virtual machines—which we strongly advocate in this book—are that users can create clean environments very quickly, and that an environment in which a bug has been found can be saved in that state for later investigation. To get these benefits, you have to install Microsoft System Center Virtual Machine Manager (SCVMM). SCVMM provides the facility of keeping virtual machines in libraries; Lab Manager lets you combine virtual machines into virtual environments in which all the machines in the environment can be stopped, started, and stored as a single entity.

Users can also create *standard environments* which can be composed of any machines. They can be physical or virtual, and if they are virtual, they don't need to be managed by SCVMM. Standard environments don't have all of the benefits of virtual environments, but they help you deploy and run tests. We recommend standard environments for some kinds of performance tests, but prefer virtual environments for most other cases.

You can set up Lab Manager without installing SCVMM or without installing Hyper-V. In that case, users can create only standard environments. This might be a useful scenario if the team has a lot of investment in testing frameworks based on physical machines or on third-party virtualization frameworks.

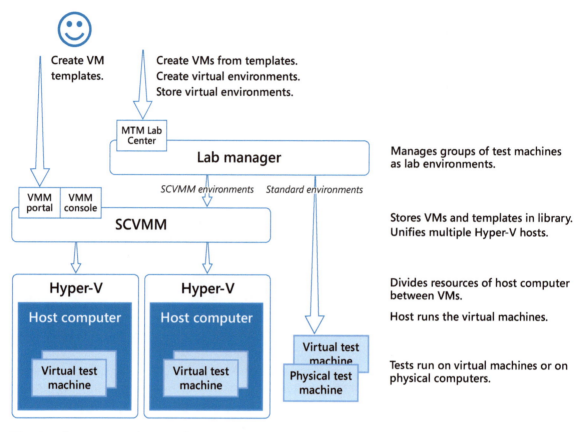

The virtual management framework

EXAMPLE CONFIGURATIONS

Now let's think about how we're going to spread the infrastructure components over different physical computers. Looking back at the diagram of components, we can draw different boundaries around them to assign the components to different boxes. As we've said, there are many arrangements that will work. Let's consider a few.

The non-starter setup

If we really want to economize, should we just put everything on one computer?

For example, perhaps we could pack all the infrastructure components into the same box, alongside Hyper-V. Well, it would work, and you can do it if you are really short of cash or space. But it is likely to perform badly in any serious project. Hyper-V is quite good at optimizing performance when all the resources are being used by VMs under its control; but it doesn't work so well if there are other services running alongside it, and taking up cycles at unpredictable times.

Another idea might be to run the infrastructure components in their own virtual machines in Hyper-V alongside the test machines. But this makes the machines containing Team Foundation Server and so on show up in the same list in SCVMM as your ephemeral test machines. Because you'll get used to creating and deleting test machines about as often as you break for refreshments, this arrangement carries the risk that you might delete your Team Foundation Server installation by mistake, thereby reducing your hard-won popularity with your colleagues.

We therefore recommend at least two computers.

The economy setup

In this setup, we put all the infrastructure on one computer, and Hyper-V for the Test VMs on another.

To really pare down the budget, we can omit SharePoint (or just use SharePoint Standard). To provide storage for Team Foundation Server, we could use SQL Server Express instead of SQL Server. However, these choices would mean we lose the project dashboards and online charts of the project's progress against tests.

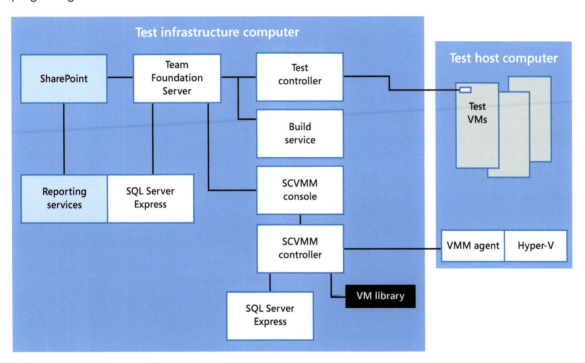

The economy test setup: two computers

In this configuration, a typical procurement spec would be:
Test host:

- 16 effective CPUs (or more)
- 48GB RAM, 64-bit architecture
- *Designed for Hyper-V* (which is not true of all 64-bit servers)
- Windows Server 2008 R2 Enterprise 64 bit
- C: drive 100GB for Windows
- E: drive 2000GB for virtual machines

Infrastructure server computer:

- 4 or more effective CPUs
- 10GB RAM or more, 64-bit architecture
- Windows Server 2008 R2 Enterprise 64 bit
- C: drive 200GB for software
- E: drive as large as possible for VM Library

Software:

- 2 x Windows Server 2008 Enterprise R2 64 bit
- Team Foundation Server 2010
- System Center Virtual Machine Manager 2008 R2 SP1
- SQL Server Express (or SQL Server 2012 or 2008 Standard)
- (SharePoint Standard)

Desktop machines for developers:

- Windows 7 Professional or Enterprise
- Visual Studio Professional (or Premium or Ultimate)

Desktop machines for testers:

- Windows 7 Professional or Enterprise
- Visual Studio Test Professional (or Ultimate)

Separate server computers for high traffic

To scale up somewhat, let's buy another three computers. By moving components out to them, we intend to reduce the impact of sudden loads in one component on users of the others. The components we move out are:

- Build Service, which has high CPU and disk load during a build.
- Test controller, which collects large amounts of data during a test run.
- SCVMM, which transfers large amounts of data when saving or restoring a VM. It's best to keep it with its library and database.

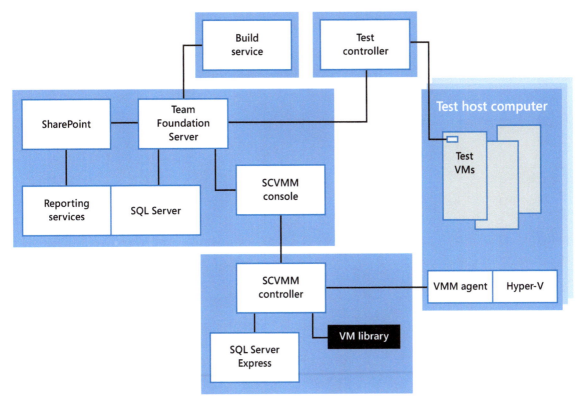

Separate build, test, and virtual library controllers

Separate file server for the VM library

In the following configuration:

- The SharePoint server is separated out, allowing it to be used heavily as a project resource, independently of the Team Foundation Server instance.
- An additional VM library runs on its own box. Additional libraries can be in separate computers, though the initial library must always be on the same box as the SCVMM controller.

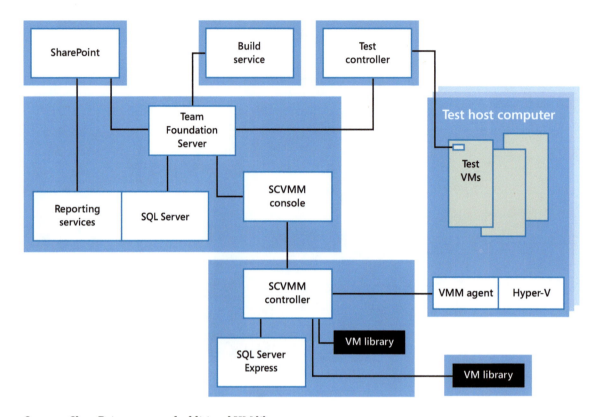

Separate SharePoint server and additional VM library

More computers

As your project grows, you might use more computers to:

- Create more test host computers. Each one will have Hyper-V and a VMM agent.
- Provide a full SQL Server installation for SCVMM, instead of using SQL Server Express. This improves performance if you have more than about 150 test machines.
- Create a Team Foundation Server cluster consisting of several Team Foundation Server instances.
- Create multiple build services and test controllers.

Server components on virtual machines

Lastly, it is possible to put all the framework components in one or more virtual machines. This makes it easy to replicate the whole setup in another project. In the following example, each virtual machine has its own (licensed) copy of Windows Server, along with one of the server components:

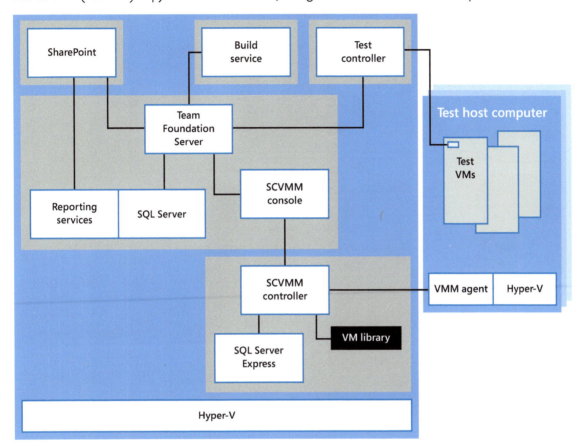

Framework components in virtual machines

This puts the server components all back into one physical computer. The only difference from our first configuration is that the components are in separate virtual machines. So what's the benefit of that?

Firstly, by putting, say, the Build Service into its own VM, you are limiting its ability to interfere with other processes; even when it is running at capacity, it can't use more CPUs or RAM than you have assigned to it, and thus cannot take power away from the SharePoint and Team Foundation Server instances that are responding to team members' queries.

Secondly, the separation provides a security barrier; a script running on the build server cannot do damage to the Team Foundation Server database.

Lastly, you can take snapshots of these virtual machines and replicate them when you want to set up a new project with its own servers.

For useful posters about configuring Team Foundation Server, download *Visual Studio ALM Quick Reference Guidance.*

Licensing

When you create virtual machines, you are creating a new copy of the Windows operating system, and also of anything else you installed on the virtual machine template. You must therefore consider how those items should be properly licensed.

Please refer to *Windows Activation in Development and Test Environments*:

To summarize the situation at the time of writing:

- Copies of Windows on virtual or physical test computers need not be activated unless you want to keep them for more than 30 days. It is also possible to extend the grace period to 120 days.
- Windows on other machines—whether physical or virtual—must be activated. This would include host servers and all infrastructure machines, such as the Team Foundation Server.

- If your organization has a Key Management Service (KMS) host, new instances of Windows on your network are automatically activated, using your organization's volume license. This is the simplest situation: if you don't have KMS, consider setting one up.

Service accounts

You will need to set up a number of service accounts in your organization's network. These accounts are used by server components to access each other and require domain access. You don't want to use personal accounts for this purpose, because you'll want to share the passwords between several administrators.

Service accounts should have no entry in your organization's email address list.

Make arrangements to update the password on a schedule. Updating a password means updating the places where the password has been entered, so that the test framework doesn't stop working.

When you are setting up most of the services, you need to have administrative permissions.

Many of the services that you need to set up have the option to run under the built-in Network Service account. Choose that where you can.

Here is a summary of services and the accounts required:

- Test Controller and agents
  - Mostly run under the built-in Network Service account.
  - To run coded UI tests (described in Chapter 5, "Automating System Tests"), the test agent has to be running as a desktop application with a user logged in. This user should be a service account.
- Build controller
  - Network Service account.
- Team Foundation Server
  - Most functions run under the built-in Network Service account.
  - Lab Management (VMM) runs as a service account.
- SQL Server
  - Network Service.
- SCVMM
  - The same service account as Lab Management.
- SharePoint
  - A separate service account.

Service account usage

The following diagram sets out the accounts that are used to access different parts of the testing framework. In this diagram, an arrow indicates that the node at the source end has to be in possession of credentials that can be used to access the target end.

> **Note:** *This diagram and the notes in this section might not make much sense until you have performed some of the installations. Come back to look at it later.*

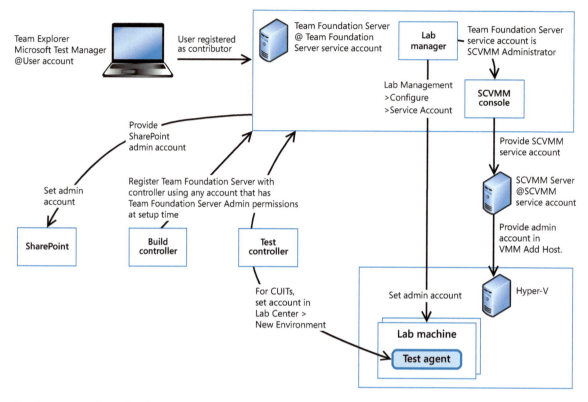

Service account dependencies

The dependencies depicted in the diagram are:

- The client machine runs under a user account. To access Team Foundation Server and its features, the user must be registered with the team project as a Contributor.
- Team Foundation Server Admin:
- You provide this account on installation.
- You need this account to connect Team Foundation Server Administrator Console to the Team Foundation Server.
- You also provide it to the build controller and test controller when you register them with Team Foundation Server.
- To manage the build controller, use Team Foundation Server Administrator Console and connect to the build controller machine.
- To manage the test controller, use the test controller management tool, which you can install from the Team Foundation Server DVD.
  - Lab Machine Admin:

- You need to add the Lab Machine Admin account to the Administrators group in every lab machine (virtual or physical). It can be a local account on each machine, but must have the same password on every machine.
- The Lab Management feature is controlled by opening Team Foundation Server Administrator Console. Connect to the team collection and open the Lab Manager page.
- In Lab Manager console, choose **Configure, Service Account** and provide credentials of the Lab Machine Admin account.
- When you set up lab machines, always add this same account to the Administrators group.
  - SCVMM Admin:
- SCVMM runs under a service account, which you provide.
- SCVMM is controlled with the SCVMM console, which should be installed in the same machine as Team Foundation Server.
- To connect to SCVMM from the console, you need the SCVMM service account password.
- Add the Team Foundation Server service account to SCVMM Administrators group.
  - Hyper-V Admin:
- On each Hyper-V host computer, you must add the Hyper-V Admin account to the Administrators group.
- You provide credentials of this account when you use the Add Host command in SCVMM console.
  - Test agent for UI tests.
- When you set up a new environment in Lab Center, you can configure one of the lab machines to run coded UI tests. You must provide an account under which to run the tests. Lab Center will configure the test agent on that machine to log into that account.

For more information about service accounts, see *Service Accounts and Dependencies in Team Foundation Server*. For Visual Studio 2010, see *Accounts Required for Installation of Team Foundation Components*.

Installing the software

OK, so at this point you have:

- Decided upon, bought, and powered up the physical hardware that you need.
- Decided which infrastructure components will go on each computer.
- The infrastructure products on DVDs (or downloaded the equivalent ISOs or installer files).
- Registered the service accounts in your corporate network.

Now, you might think that it's unnecessary for us to include here any long-winded advice about how to install some software. You are looking forward to a relaxing day of reading your favorite testing blogs while intermittently changing a DVD, clicking through an installer wizard, and mindlessly accepting obscure legal terms and conditions. Furthermore, there is plenty of help on the web of the "Step 43: Click OK" variety, should you feel the need to have your hand held through the process.

However, you might find that there are a few options presented by the installers that feel a bit like "Would you prefer to exist in a single infinite universe, or in a finite brane in an infinite multiverse? This choice cannot be undone, and is final and complete and forever, and must be made now." In addition, all the help about the different components are in different places, and are generally reluctant to acknowledge each other's existence.

We therefore feel it might be useful to offer a few observations and recommendations that are aimed specifically at integrating the components into a testing infrastructure.

INSTALLATION OVERVIEW

Follow these steps to set up your test infrastructure. The ordering is important, because of dependencies between the parts:

1. **Set up Hyper-V** to provide a host computer for test virtual machines.
2. **Set up computers** for test infrastructure products.
3. **Install System Center Virtual Machine Manager (SCVMM)** for virtual environments.
4. **Install SQL Server 2008 Standard** for project management charts and reports showing burndown. (You could use SQL Express, but you would not get the most useful project progress reports.)
5. **Install SharePoint** for the project portal and reports on test progress.
6. **Install Team Foundation Server**:
    a. Configure the Build Service.
    b. Configure Lab Management.
7. **Populate the virtual machine library** with a base set of machines that will be used for tests. Users can replicate and modify the library contents, so the first step is just to provide machines that contain the principal operating systems that they are likely to use.

HYPER-V

Configure Hyper-V on your test host computer. If you have decided to run your infrastructure components in virtual machines, you will also want to install Hyper-V on a separate computer.

A machine on which you want to install Hyper-V should have: 64-bit architecture; hardware-assisted virtualization; Windows Server (we use 2008 R2); at least one very big disk; lots of RAM; and many logical CPUs. It must be a physical computer. We recommend you don't use any other role. (See *Install the Hyper-V Role on a Server Core Installation of Windows Server 2008* on MSDN.

To configure Hyper-V:

1. Make sure the computer is joined to your network domain.
    a. In the **System** control panel, look under **Computer Name**.
    b. Choose a name for this computer that ends in "-HOST". This helps avoid confusion between virtual and physical machines.
2. In **Server Manager**, choose **Roles**, **Add Roles**, **Hyper-V**.

Enabling Hyper-V

3. Set the default storage locations to this computer's very big disk:
   a. In Hyper-V Manager, select this computer and choose **Hyper-V Options**.
   b. On the **Virtual Hard Disks** page, enter a folder such as D:\VHD.
   c. On the **Virtual Machines** page, enter a folder such as D:\VM\.

Virtual machines for test infrastructure components

This section is about installing test infrastructure products (Team Foundation Server and so on) on virtual machines. If you're going to install those products on bare metal, skip this section.

> **Note:** *This section isn't about creating the virtual machines you will use for testing. Once SCVMM is running, you will use that to set up test VMs. We'll discuss that later.*

There are two methods for setting up virtual machines:

- Install Windows from a PXE server, installer disk, or ISO file, as you would with a physical computer (see *Walkthrough: Deploy an Image by Using PXE*; or
- Replicate an existing VM that already has Windows installed.

Your first virtual machine: installing Windows on a new blank VM (60 minutes)

1. **Create the VM**. In Hyper-V Manager, choose **Action**, **New**, **Virtual Machine**. Enter these choices:
   - **Specify Name and Location**:
     - **Name**: Use the same name as you will use for the domain name.
     - **Store**: Make sure it is on the big disk drive.
   - **Configure Networking**: choose **Local Area Connection – Virtual Network**.
   - **Connect Virtual Hard Disk**:
     - Location should be on the big disk drive, for example D:\VM\.
     - You can add more hard disks later.
   - **Installation options**: If your organization has a PXE server, choose **Install an operating system from a network-based installation server**. Otherwise, select your Windows DVD or .ISO file, which you should make available to the host machine.
   - When you finish the wizard, the VM will appear in the list of machines, in the Off state.

2. **Start the VM** and choose **Connect**. The VM's console screen will appear.

   If you selected a network installation, you might need to press F12 to choose the operating system.

   Windows will be installed. When asked, provide an Administrator password.

   If your network doesn't have KMS, you'll also have to provide a Windows activation key. (See the *Volume Activation* topic on TechNet.)

3. **Log into the console**. To log into a machine in Hyper-V, choose Connect to open its console window, and then choose the CTRL+ALT+DEL button at the top left of the console window.

Logging in

4. **Perform housekeeping**:

   a. **Get updates**. In the Windows Update control panel of the VM, set updates to Automatic. You will typically have to restart to complete the updates.

   b. Enable Remote Desktop.

5. **Add processors and disks**:

   a. Shut down the virtual machine.

   b. In the Hyper-V console, select the VM and choose **Settings**.

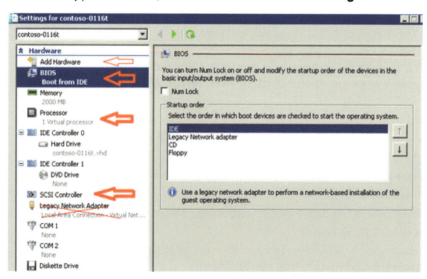

Virtual machine settings

   c. On the **BIOS** tab, move IDE to the top of the Startup order list. This prevents the VM from trying to start from PXE in the future.

   d. On the **Processor** tab, adjust the number of logical (= virtual) processors.

   e. On the **SCSI Controller** tab, you can add extra virtual hard drives.

       Specify a fixed-size disk. Don't forget to set its location on the big disk of the host.

   f. Remove the **Legacy Network Adapter**.

       Choose **Add Hardware**, and then **Network Adapter**. This type of adapter runs more efficiently.

   g. Save the settings and start the VM.

   h. If you added a disk, you will need to mount it in the VM: Choose **Administrative Tools**, **Computer Management**, **Storage**, **Disk Management**. Select the new disk and **Initialize** it. Then create a **New Simple Volume** on it.

6. **Export the VM** so that you can use it as a template to create new virtual machines:

   a. Install any additional software that you want to be replicated to every copy.

   b. **Shut down** the VM.

   c. In Hyper-V Manager, select the VM and choose **Export**. Export to a separate folder on the big disk such as D:\exports\.

       The export operation can take 5-20 minutes, and runs in the background. You'll know it has finished when the **Cancel Exporting** command disappears.

7. **Delete the VM**. You are going to use the exported version as a template from which to create copies. Deleting the original helps you avoid name clashes when the copies are started.

   a. Delete the VM in Hyper-V.

   b. Delete the disk files on the host in D:\VHD\.

More virtual machines: creating a virtual machine by import (15 minutes)

Make sure the original VM— from which you exported—is paused or shut down before you create a VM by import. The new VM will have the same domain name as the one from which it is copied. You must therefore change the domain name before you can have both machines running.

1. In Hyper-V Manager, choose **Import Virtual Machine**.

2. Select these options in the wizard:

   • **Copy the virtual machine (create a new unique ID)**.

   • **Duplicate all files so the same virtual machine can be imported again**.

   The operation typically takes 3-10 minutes.

3. Rename the new machine in Hyper-V. This changes only the name by which it is identified in Hyper-V.

4. In D:\VHD, rename the new computer's virtual disk files:

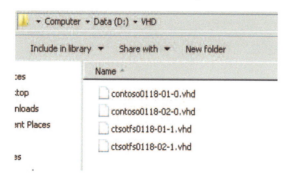

Virtual hard disk names

5. In Hyper-V, open the **Settings** of the new machine, and edit the definition of each disk. Change the location of the disk to point to the new virtual disk file names.

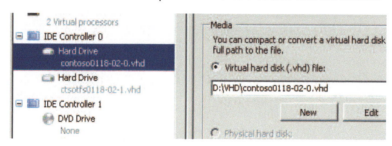

Change location

6. Start the new VM and log into it.

7. **Join the computer to a domain** by using the System control panel of the VM. Restart the machine.
   Make the computer's domain name the same as its name in Hyper-V; this isn't automatically the case.

8. **Add users to the Administrators group**. Choose **Administrative Tools**, **Computer Management**, **Local users and groups**, **Groups**, **Administrators** and add aliases to the group. Don't forget yourself.
   From this point, you can log into the VM by using Remote Desktop.

INSTALL SYSTEM CENTER VIRTUAL MACHINE MANAGER (SCVMM)

SCVMM lets you keep a library of virtual machines. It also allows a whole set of Hyper-V hosts to be treated as though they were a single pooled resource. SCVMM is a pretty powerful tool, allowing you to create your own local cloud. It can even move VMs between one server and another while they're working to optimize performance.

We'll use SCVMM to supervise test VMs.

Prerequisites:
- An account on your domain: the VMM Service Account.
- A domain-joined physical or virtual x64 machine with Windows Server 2008 R2 SP1. This can be the same computer as a Hyper-V test host.
- A large disk on which to store the VM library.
- The installation DVD and product registration key for System Center Virtual Machine Manager 2008 R2.

Install the VMM server
1. Using Remote Desktop, log into the virtual or physical machine on which you plan to install SCVMM.
2. Start the SCVMM installer, and then choose to set up **VMM Server.**
3. When the wizard asks whether to install SQL Server Express, say yes. (But if you plan to run more than 150 hosts for test virtual machines, say no, and install SQL Server 2008 on a separate machine.)
4. The Library Share should be on the large disk. You can add another share on another machine later if required.
5. Change the default VMM communication port from 80 to another value such as 8000. This leaves port 80 free so that you can install SharePoint or the Self-Service Portal on this machine.
6. Provide the VMM Service Account.

Install other bits
Use the same installer disk or ISO to install the following pieces:

Component	Where	What for
VMM administrator console	VMM server machine	Aids administration.
	Team Foundation Server machine	Enables lab management.
	Your desktop	Aids administration.
VMM local agent	Test Host Hyper-V computers; but the VMM server will normally install this for you.	Allows SCVMM to control hosts.
VMM self-service portal	VMM server machine	Allows users to create VMs.

When you install the VMM Self-Service Portal, you need to:
- Configure it to use a port other than 80; or
- Place it on a separate machine from SharePoint

Connect to Hyper-V
In the SCVMM console, use the **Add Host** command, and provide credentials for a service account that has administrator permissions on the Test Host box. SCVMM will add this service account to the Virtual Machine Manager security group on the host machine. It will be used to control the SCVMM agent.

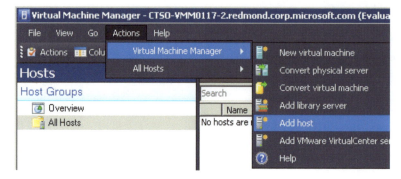

Registering a Hyper-V host with SCVMM

This will install the VMM agent on the Hyper-V computer, and start the Hyper-V role if necessary. It can take a few minutes.

Enable Microsoft .NET 3.5 and Internet Information Services (IIS)

In Server Manager, open **Features**, and enable **.NET 3.5**

In Server Manager, open **Roles**, and add **Web Server (IIS)**. Enable its role services:

- Common HTTP Features
- Application Development
- Management Tools\IIS Management Console

See *How to: Configure Team Foundation Server By Using the Standard Configuration for Single-Server Installations* on MSDN.

Install SQL Server 2012 or 2008 R2 for Team Foundation Server

You have to install SQL Server 2012 or 2008 R2 if you want project management charts and reports showing historical data such as burndown. We recommend this. (If you don't install one of these versions at this point, the Team Foundation Server basic installer will install SQL Server Express instead.)

1. In the SQL Server installer, select the following features:
   - **Database Engine Services\Full-Text Search**
   - **Reporting Services (each variant)**
   - **Integration Services**
2. On the **Instance Configuration** page, set **Default** instance.

   (If you set any other instance name, you will have to use **Advanced** setup in Team Foundation Server, and refer to the database as machineName\instanceName.)

3. On the **Server Configuration page:**
   - **Provide your SQL Server service account credentials. Click Use the same account for all.**
   - **Set the Startup type of SQL Server Agent to Automatic.**

4. On the **Database Engine Configuration page**:
   • **Choose Add Current User.**
   • **Add your Team Foundation Server service account.**
5. On **Reporting Services Configuration**
   • **Choose Integrated Mode if you will be installing SharePoint.**
   • **Choose Native otherwise.**

Enable TCP and Pipes protocols
When the installation is complete, open SQL Server Configuration Manager.

Open **SQL Server Network Configuration/Protocols for MSSQLSERVER**. Make sure that the **Named Pipes** and **TCP/IP** options are enabled. If you have to change them, open the **SQL Server Services** page and stop and restart the **SQL Server** service.

Install SharePoint (if required)

You can skip this step if you don't expect to make heavy use of SharePoint. The standard installer for Team Foundation Server includes an installation of SharePoint.

SharePoint provides an enhanced project website, including dashboards that show collections of reports on the project's progress. It also provides a project portal where you can post announcements and share documents. (See *Manually Installing SharePoint Products for Team Foundation Server* on MSDN.) If you expect only to use the dashboards and share a small number of documents, skip this step.

However, if your team will make serious use of SharePoint, we recommend that you install it on a separate machine from Team Foundation Server. It works best on a machine with at least 8GB RAM.

You might also install SharePoint now if you want to use the more advanced installer for Team Foundation Server, which does not include SharePoint installation; or if you want to install SharePoint Enterprise. By using the advanced Team Foundation installer, you can also integrate it with an existing SharePoint site.

Once you have installed SharePoint, test it with your web browser: http://servername/.

To administer SharePoint, note the URI provided by the installer; for example, http://servername:17012.

Install Team Foundation Server

Run the Team Foundation Server installer from the DVD. It installs and configures prerequisites such as .NET Framework 4.5, and might require you to restart the computer. If so, the installer will resume when you log in again.

Most of your choices are made in the Configuration Manager, which starts up when the installer is finished.

Configure Team Foundation Application Server
The initial choices in the configuration wizard are:

Standard Single Server – Choose this if you are installing SharePoint and SQL Server on the same machine as Team Foundation Server, and if your SQL Server database has the default name. If you have not already installed SharePoint, this wizard will install it for you.

Advanced – Typical reasons for using this are:

- Your SQL Server or SharePoint Server installations are on separate machines.
- You gave your SQL Server a name other than the default, in which case you'll have to enter it as machineName\serverInstanceName.
- You want to set up a cluster of Team Foundation Server machines that will function as a single large-scale service.
- Any of the assumptions made by the Standard Single Server wizard about how you've set up the rest of your kit turn out to be incorrect.

Basic – This doesn't give you reporting or a SharePoint portal. But you probably want both of these, because you want to be able to see charts that show how many tests are passing for each of your user stories.

Enter your Team Foundation Server service account in the wizard. Note that the Test button checks only that the credentials are valid in the domain; it doesn't check that all the relevant services can be accessed.

When the configuration has finished, note the connection details:

Completed configuration

The Web Access URI can be used from a web browser.

The Team Foundation Server URI is entered into Visual Studio or Microsoft Test Manager in the Connect to Server dialog.

Set up Lab Management

1. Start Team Foundation Server Administrator Console.

2. Expand **Application Tier**, **Lab Management**.

3. Choose **Configure Lab Management**.

4. If prompted, provide your own account (or an Admin account on the SCVMM machine). In the wizard, the default **Network Isolation** and **Advanced** settings are OK.

Create a team project collection

In the Team Foundation Server Administrator Console, in **Team Project Collections**, choose **Create Collection**.

1. On the **Data Tier page,** give the machine the name of the SQL Server 2008 instance. Unless you created a separate SQL Server instance for Team Foundation Server, this will just be the local machine name. Choose **Create a new database**.

2. On the **SharePoint Site** page, Team Foundation Server will create a SharePoint sub-site as this project collection's portal. In the Web application, you need the web address of the SharePoint site; if you put everything on the same machine, it will just be http://thisMachine. Make a note of the URLs of the project portal sites.

3. On the Reports page, Team Foundation Server will set up a SharePoint sub-site on which you can view generated reports. Make a note of the URLs.

Set up the build service

You can set up the build service on the same machine where Team Foundation Server is installed, or on a different machine. Using a separate machine avoids any sudden slowdowns in Team Foundation Server when a build starts. (See *Scenario: Installing Team Foundation Build Service* on MSDN.)

> **Note:** *Don't forget to install any platform software on the build server so that your built software can run on it. For example, if you are developing software that uses .NET and SQL Server, install the correct versions of each product.*

Install the build service from the same DVD or ISO (disk image) as Team Foundation Server.

In the Team Foundation Server Administrator Console on the build service machine, select **Build Configuration**, and choose **Configure Team Foundation Build Service**. The main item to choose is the path of the Team Foundation Server project collection; the default settings usually work for everything else.

The wizard sets up a build controller and agents.

If you set up the first team project collection after you start the build service, you will have to come back to the Build Configuration console. Open Properties to connect the service to a team project collection. You can then use the New Controller and New Agent commands, choosing default values for everything.

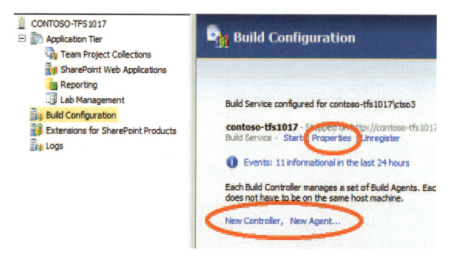

Build configuration

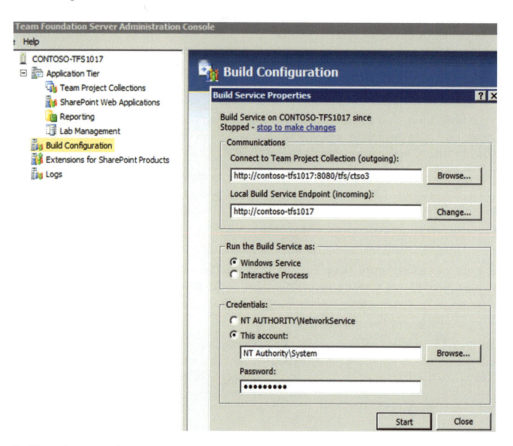

Build service properties

Allowing build completion emails

Open Team Foundation Server Administrator Console, and under your server machine name, select **Application Tier**.

Make sure that **Service Account** is set to an account that has permission on your domain to send email messages. If it isn't, use **Change Account**. Don't use your own account name because you will change your password from time to time. (If you ever have to change the password on the service account, notice that there's an Update Password command. This doesn't change the password on that account; it changes the password that Team Foundation Server presents when it tries to use that account.)

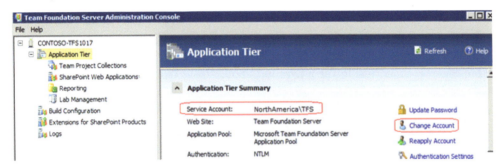

Setting the email notification account

Scroll down to **Email Alert Settings** and choose **Alert Settings**. At this point, you'll need the address of an SMTP server in your company's network. Get one from your domain administrator.

Email alert settings

CONNECT TO TEAM FOUNDATION SERVER IN VISUAL STUDIO

From your desktop computer, you should now be able to connect to Team Foundation Server:

Connecting to your server in Visual Studio

Create a project in the new collection. In Team Explorer, choose **New Team Project** on the shortcut menu of the team collection.

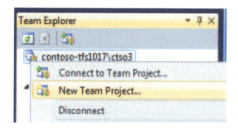

New team project

ADD TEAM MEMBERS

You can control access by individuals and by teams. Typically you create a team in Team Foundation, and then set the permissions of the team. Individuals can be added or removed from the team.

In Visual Studio, choose **Team**, **Team Project Settings**, **Group Membership**. Add people in your team to the default team. Also consider adding individuals to the **Project Administrators** group.

CONNECT IN MTM

Connect to your project from Microsoft Test Manager:

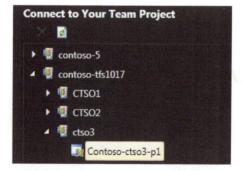

Connecting to your project in MTM

SETTING UP THE INFRASTRUCTURE   221

SET UP A TEST CONTROLLER

A test controller is used to coordinate the deployment, running, and monitoring of tests on remote machines. If you want to run tests on any machine except the one where you built the software, you need a test controller. In particular, you need a test controller to run tests of client/server or other distributed systems.

One place to set up a test controller is on any machine—for example, your own desktop—from which you want to coordinate tests. In this case, you use Visual Studio as the user interface to the test controller.

Alternatively, you can register a controller with a team project collection. You need to do this if you want to run on test environments. You can install one test controller on any computer. The easiest location is the same machine on which the team project collection is located.

In your Team Foundation Server installation pack, find and run the installer for test controllers. You might also be able to find an ISO file that can be downloaded from the web by searching on "Visual Studio Agents."

To install from an ISO file onto a virtual machine:

1. Copy the ISO file to the Hyper-V computer that hosts the virtual machine on which you want to install the controller.

2. Use Hyper-V to pause the installation machine. Open its settings and on the DVD page, mount the ISO file on the installation machine's virtual DVD drive.

3. Resume the installation machine.

4. Log into that machine and install the test controller from the DVD.

When the installation has finished, check the **Configure Now** box.

If you are creating the test controller to run test cases from Microsoft Test Manager, configure the test controller as a Network Service, and register it with one of your Team Foundation Server collections.

Alternatively, if you are creating a standalone test controller to run through Visual Studio, set it up on your Visual Studio machine. Both the test controller itself and the test controller management tool are installed. (In Visual Studio 2010, use the Manage Test Controller item on the Test menu.)

Setting up physical and virtual machines for testing

You've now set up the basic testing infrastructure. One thing remains to be done, which is to populate your library with virtual machine templates that you can use for testing. You might also want to configure some physical machines for testing.

Typically you'll set up two or three virtual machines with different operating systems from scratch, and after that you can create the others by making copies in which you adjust the configurations.

You are likely to need at least four templates:

- **Test lab servers**. These are machines on which a tester will typically install the server part of a client/server system.
- **Manual testing client**. Testers will use this template to test web clients or desktop applications. See Chapter 4, "Manual System Tests." It includes:
    - Web browsers – several makes and models, if you are testing a web application.
    - Visual Studio Test Professional. This provides you with the Test Runner, which helps a tester work through the tests and can record and play back their actions.
- **Automated test development client**. VMs created from this template are used to debug automated tests. See Chapter 5, "Automating System Tests." It includes:
    - Web browsers.
    - Visual Studio Ultimate.
- **Domain Controller**. This VM is used to provide a domain name server in a network-isolated environment. For more information, see *Working with Network Isolated Environments* in Chapter 3, "Lab Environments."

CREATING A VIRTUAL MACHINE BY USING SCVMM

The first virtual machines in your library have to be created by using SCVMM. You can then store them in the SCVMM library, and import them into Lab Manager. From there, users can adapt the initial virtual machines and store new versions in the library.

> **Tip:** *Configure the SCVMM Self-Service website. This allows team members to create new virtual machines without having administrative privileges on the SCVMM server.*

Follow the instructions in the MSDN Topic: *How To: Create and Store virtual machines and templates ready for Lab Management*. In summary the process is as follows:

- Create a new virtual machine and install Windows. If your organization has a PXE server, it makes the installation easier.
- Add a user account that has Administrator privileges and is the same on every lab machine. This allows Lab Manager to administer the machine.
- Install a test agent from the Team Foundation Server DVD; but, do not connect it to a test controller yet. (Lab Manager will do this when you use the machine in a lab environment.)
- Configure Windows on this machine as you will want it on machines that are created from this template. For example, you might want to enable the Web Server (IIS) role.
- Install whatever other software you want to exist on machines created from this template; for example, Visual Studio.
- Clear the Administrator password and the local password policy. These are special requirements for saving the machine as a template.
- Shut down the machine.
- In the SCVMM console, store the machine in the library as a template.

Where to go for more information

All links in this book are accessible from the book's online bibliography available on MSDN: *http://msdn.microsoft.com/en-us/library/jj159339.aspx.*

Index

www.ingramcontent.com/pod-product-compliance
Lightning Source LLC
Chambersburg PA
CBHW041416050326
40689CB00002B/540